Industrial Robotics

Industrial Robotics

Selection, Design, and Maintenance

Harry Colestock

McGraw-Hill

New York   Chicago   San Francisco   Lisbon   London   Madrid
Mexico City   Milan   New Delhi   San Juan   Seoul
Singapore   Sydney   Toronto

The McGraw·Hill Companies

Cataloging-in-Publication Data is on file with the Library of Congress.

Colestock, Harry.
    Industrial robotics : selection, design, and maintenance / Harry Colestock.
      p. cm.
    ISBN: 0-07-144052-6
    1. Robots, Industrial.   2. Robotics.   I. Title.

    TS191.8.C65   2004
    670.42'72—dc22

                                          2004055194

1 2 3 4 5 6 7 8 9 0 DOC/DOC 0 1 0 9 8 7 6 5

ISBN 0-07-144052-6

The sponsoring editor for this book was Judy Bass and the production supervisor was Sherri Souffrance. It was set in Sabon by Patricia Wallenburg. The art director for the cover was Anthony Landi.

Printed and bound by RR Donnelley.

 This book was printed on recycled, acid-free paper containing a minimum of 50% recycled, de-inked fiber.

McGraw-Hill books are available at special quantity discounts to use as premiums and sales promotions, or for use in corporate training programs. For more information, please write to the Director of Special Sales, McGraw-Hill Professional, Two Penn Plaza, New York, NY 10121-2298. Or contact your local bookstore.

Contents

Preface

With the rapid proliferation of many types of robots in recent years, a need has been recognized to assist the potential robot user in the proper selection, care, and feeding of his or her robot to achieve the maximum in economic productivity. This book is intended to describe the various factors that influence the application of robots to blue collar tasks. Comparisons are made with "fixed automation" assigned to similar tasks so that better judgments can be made in their application.

As with any new technology, communications and understanding suffer until a good vocabulary is established and definitions of terms have been made. An initial attempt is made to classify robots in accordance with their complexity and functions provided.

Over the past 10 years, significant advances have been made in robot capability in the areas of operating work volumes, and repeatability accuracy. Also, prices have continued to drop. A typical example might be that a $100,000 robot of 10 years ago might be purchased today for around $30,000.

One of the brightest factors influencing the application of robots has been the introduction of machine vision. Machine vision has opened doors for robot applications that never existed only a few years ago. It is normally superior to manual inspection because it can handle high speed and volume production; it offers more precise readings and provides better process control. Unlike a human operator, machine vision systems can not only report, document, and inspect for failures, but can also provide the reason for and the extent of that failure. Machine vision's ability to avoid false "accepts and rejects" is especially important to manufacturers who are cost-conscious to maximize throughput with minimal waste.

A typical example of the way the industry is heading was reported in the October 10, 2003 Detroit Free Press. Nissan's new SUV assembly line in Canton, Mississippi has 17 miles of conveyor belts, 853 robots, 70 lasers, and 300 programmable logic controllers that control the entire assembly process. This $1.43 billion plant has a dire need for maintenance technicians to keep the expensive machinery running; training programs to provide at least 200 key technicians have been instituted.

x

Acknowledgments

The author would like to thank his wife, Marilyn, for the typing of the rough draft and the several rewrites and additions to the text. He would also like to thank Jeff Donovan, Chief Engineer, Robotics, of Cinetic Automation and its president, John Neuman for their support and information files on many of the cases illustrated that compared robotics with fixed automation. Let it be explained that much of the author's experience in automation and robot applications was obtained while serving as Director of Engineering of the Automation Division of Ingersoll-Rand which was subsequently purchased by Cinetic Automation, a French firm.

Next, I would like to recognize the two leading robot manufacturers, ABB Automation Inc. and FANUC Robotics North America, both from Michigan, for their generous offering of modern-day examples and case histories of robot applications.

Last, but not least, I would like to thank my wife, Marilyn, for allowing me to take time away from my fine hobbies, gardening, magic, watercolor painting, writing poetry, and medical engineering, plus many of my household chores, to write this book hopefully as a help to those who may need a robot to improve their productivity and thereby their profitability.

Robots—Definitions and Classifications

Definitions

The industrial robot, as presently conceived and utilized, is an automatic programmable transfer and handling machine. In contrast to conventional forms of automatic transfer and loading devices, robots can be designed so that they handle a plurality of entirely different work pieces. They can be programmed to carry out a great number of different movements and operate with the speed and efficiency of automatic special purpose machines.[1]

Webster's New Twentieth Century Dictionary lists a robot as:

1. An automaton manufactured by Rossum in Carel Kapek's play R.U.R. (Rossum's Universal Robots) having immense strength, but no feelings; hence, an efficient worker devoid of feelings.
2. A mechanical contrivance which performs functions with almost human intelligence, particularly one operated by light rays or radiant energy; for demonstration purposes a robot is usually made to resemble a mechanical man, which walks, talks, and operates machines, but for practical purposes the robot may be merely a device attached to a piece of machinery to increase its efficiency.[2]

For the purpose of our discussion here, we are adding one word to the former definition. A robot is defined as an automatic articulated programmable transfer and handling machine. This will differentiate between a robot and a computer-controlled transfer line.

Robot Classifications

Because robots have been utilized to replace humans on the production line, we tend to agree with Dr. Ghali in his classification of robots according to their intelligence.[3] Intelligence has been defined as the ability to utilize experience in adapting to new situations.[4] Therefore, most of the blue-collar robots available today would not classify as having intelligence. However, both Dr. Ghali and Dr. Leo C. Driscoll are predicting these self-adapting types for production line work within the next few years.

The Superintelligent Robot

This robot is presently under development. It is able to reproduce the movements of human legs and arms. With a built-in brain, it will be able to make decisions and very probably will be used to assemble parts. Its production costs could exceed one-half million dollars. These machines will not only duplicate the movements of an operator in an exact manner, but will, with their built-in brain, have the ability to work without an operator.

The Intelligent Robot

This robot will possess machine vision, a computer, and sensors. The movements take place according to a defined program, as well as from instructions arriving from the sensors and the television camera, which registers objects moving in front of it. These robots are presently opening up opportunities in industry that were not anticipated only a few years ago (see chapter on *The Impact of Machine Vision on the Robotics Industry*).

The Nonintelligent Robot

Dr. Ghali divides this class into three categories:

The Universal Robot

This robot was introduced in the U.S. market in 1960. The best known are the Unimate made by Unimation Inc., and the Versatran made by

American Machine and Foundry Co. These robots possess an electric or electronic control. The arms move in three axes: vertical, horizontal, and rotation. The movements can be from point-to-point or on a continuous path and can be carried out simultaneously. The maximum load on the arms is 40 kg and the repeatability lies between 1.25 and 2.00 mm. The price for one unit is between $20,000 and $30,000. These robots are built for a life of 40,000 hours.

The Simple Robot

This robot is simpler in design than the universal robot. It is designed for application in a certain area, but is flexible without being universal. Its price lies around $8,000 to $15,000 depending on the size and complexity desired.

The Miniature Robot

These devices are used for assembling small parts and sell for about $3,000. Some are programmed by means of cams with vacuum pick-up fingers. Part handling capability is usually just a few ounces.

Fixed Automation

First, note that all classes of robots form a subgroup within the overall definition of automation. Fixed automation, as opposed to robots, however, is distinguished here to show those characteristics that give it advantages and disadvantages in different applications. Fixed automation, as the name implies, is less flexible than the robot, made up of transfer lines, index tables, and special machines. It does not try to emulate the movements of the human arm or hand. It is, in many cases, automatic and programmable, and transfers and handles parts. The main distinction between robots and fixed automation is the articulated arm and hand movements made by the industrial robot, which can be readily reprogrammed for part change or process alteration. This great flexibility is seldom available in fixed automation. Reprogramming of a computer-controlled assembly line is usually a much longer and involved process.

Fixed automation may also be characterized as being tailored to one or a limited number of parts to be handled. This limitation usually carries with it an inherent advantage of greater production speed. Also, production speed is usually greatly enhanced by several movements and processes taking place simultaneously, such as in multistation assembly machines and indexing transfer lines.

These fixed automation lines usually require very large capital investments and hence, must have adequate volume requirements to justify such expenditures. This factor, probably more than any other, has restricted the use of fixed automation where it is otherwise justified. Robot design has tried to take advantage of this fact and is usually available at a fraction of the cost of fixed automation. Care should be exercised in comparing robots and fixed automation because there often is a great difference in production rates.

One other singular advantage of an industrial robot over fixed automation is the ease with which the robot may be moved to a new location. This feature lends itself to better plant utilization and a greater flexibility to meet changing work requirements.

References

1. Ghali B. Robotology and an Overview of the International Robot Situation. *Proceedings of Second International Symposium on Industrial Robots*. May 1972:31.
2. *Webster's New Twentieth Century Dictionary of the English Language*. 1948:1472.
3. Ghali B. Robotology and an Overview of the International Robot Situation. *Proceedings of Second International Symposium on Industrial Robots*. May 1972:4–5.
4. Comparative Psychology. In: *Encyclopedia Britannica*. 1972, XVIII:774.

Economic Productivity

Economic productivity has been defined as the ratio of the output of a good or service, or collection of goods or services, to the input of one or more of the factors producing it. This ratio may be in the form of an average, expressing the total output of some category of goods divided by the total input of a factor or factors. Or, it may be incremental, expressing the ratio of a change in output to the associated change in input.[1]

Classically, it is defined as the output per man-hour or unit of input. This definition leaves something to be desired because it doesn't take into consideration such related factors as increased efficiencies in management, marketing, distribution, sales, or reduction in paperwork and menial tasks via the computer. Discussing definitions at a luncheon speech, Dr. Carl H. Madden, chief economist of the U.S. Chamber of Commerce, suggested that productivity be considered useful output with relation to total input. Related considerations such as ecology, environmental conditions, standards for noise abatement, and so on, would all be considered in the equation.[2]

Long-term trends in national output per worker are depicted in Table 2.1. Because of difficulties inherent in productivity measurement, the figures should only be taken as rough approximations; however, they still suffice for distinguishing the rates of productivity increase in four countries.

While the United states has lagged in productivity improvements behind its nearest competition, it had emerged from World War II and the Space Race with the most advanced technology in the world. It is only now beginning to feel the effects of its closest competitors, Japan and West Germany, for world markets, especially in electronics and automobile manufacturing.[3]

Table 2.1   *Long-Term Trends in National Output per Worker**
(Index Numbers, 1890 = 100)

Year	Country			
	United States	Great Britain	Germany	Japan
1890	100	100	100	100
1900	122	107	100	144
1910	138	110	107	166
1920	142	100	—	228
1929	172	116	90	366
1938	182	132	127	547
1948	223	132	—	314
1960	295	161	166	747

*Economic Productivity. In: *Encyclopedia Britannica*, 1972, VII:932.

Tables 2.2–2.7 illustrate the changing nature of labor productivity in both the United States and in countries around the world. These trends have caused alarm in the minds of many U.S. managers and were instrumental in establishing the first Manufacturing Productivity Conference held in Washington, D.C., on October 11–13, 1972.[4] Before this situation can be called a crisis, however, analysis must be done to determine if these countries with the faster growth rate in productivity are merely starting from a less advanced point in their development and are catching up to the United States. The real payoff is not so much what the growth rate has been, but what the actual productivity is at present. How much in goods or services can be achieved for an invested man-hour? Table 2.7 gives some insight into this and shows how far the other countries had to go in 1960 to catch the United States in productivity. Growth in productivity is important, however, if the United States is to maintain its leadership role in world competition. Even though we may still lead the world in output per worker, we currently pay that worker the highest wage rate in the world. Stiff competition in world markets is evidenced in the fact that since 1970, U.S. imports have exceeded exports. Also, for the first time since records have been kept, the United States dropped out of first place (to fourth) in the installation of new machine tools.[5]

Table 2.2 *Labor Productivity Manufacturing in Industry World Community (1955–1968)*

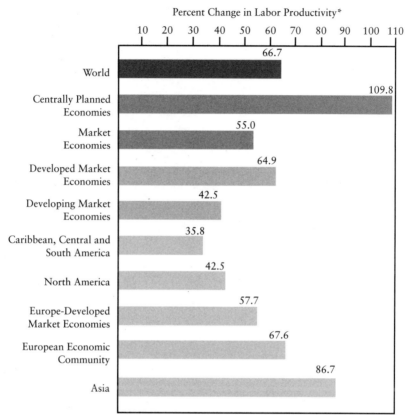

Percent Change in Labor Productivity*

World	66.7
Centrally Planned Economies	109.8
Market Economies	55.0
Developed Market Economies	64.9
Developing Market Economies	42.5
Caribbean, Central and South America	35.8
North America	42.5
Europe-Developed Market Economies	57.7
European Economic Community	67.6
Asia	86.7

* Calculated from increase of labor productivity based on
indexes of industrial production and industrial employment.
Source of data: Table 11, Index Numbers of Labor Productivity
in Industry, World, Total Ensemble 3. *Statistical Yearbook*, 1970.
United Nations, 22nd issue, New York, 1971.

7

Table 2.3 *Productivity Improvement for Selected Countries (1963–1968)*

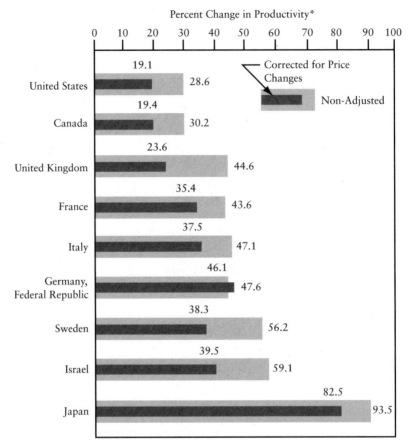

Percent Change in Productivity*

* Productivity defined as value added by manufacture per unit
of labor employed. Shown both raw and corrected for wholesale
price changes.
Source of data: Table 80, Output and Employment in Manufacturing
and Table 176, Wholesale Prices. *Statistical Yearbook,* 1970.
United Nations, 22nd issue, New York, 1971.

8

Table 2.4 *Economic Growth Rates (1950–1960 and 1960–1968) for Selected Countries*

Average Annual Percent Change in GNP*

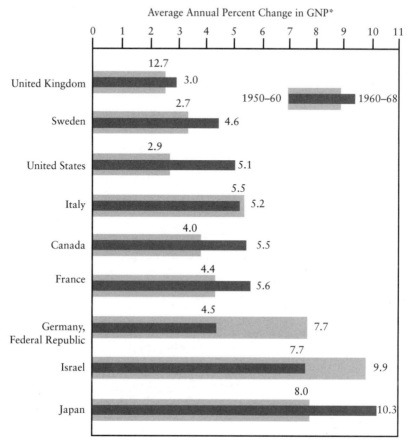

United Kingdom: 12.7 / 3.0
Sweden: 2.7 / 4.6
United States: 2.9 / 5.1
Italy: 5.5 / 5.2
Canada: 4.0 / 5.5
France: 4.4 / 5.6
Germany, Federal Republic: 4.5 / 7.7
Israel: 7.7 / 9.9
Japan: 8.0 / 10.3

1950–60 / 1960–68

* Gross national product.
Source of data: Table 181, National Accounts. *Statistical Yearbook,* 1970. United Nations, 22nd issue, New York, 1971.

9

Table 2.5 *Productivity in the U.S. Nonfarm Industries (1950–1970)*

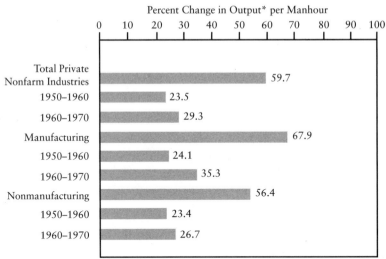

* Refers to gross national product in 1958 prices.
Source of data: Table 350, Indexes of Output per Manhour,
Hourly Compensation and Unit Labor Costs in the Private Economy:
1950 to 1970. *Statistical Yearbook*, 1970. United Nations,
22nd issue, New York, 1971.

Table 2.6 *Indexes of Output per Man-Hour—Production Workers and All Employees, Selected Industries (1950–1969)*

[1967 = 100. Prior to 1960, excludes Alaska and Hawaii. See *Historical Statistics, Colonial Times to 1957*, series W 13–3 and W 44, for indexes on a 1947 base]

INDUSTRY	OUTPUT PER PRODUCTION WORKER MAN-HOUR					OUTPUT PER ALL EMPLOYEE MAN-HOUR				
	1950	1960	1965	1968	1969 (prel.)	1950	1960	1965	1968	1969 (prel.)
Manufacturing: [1]										
Aluminum rolling and drawing	(NA)	60.2	95.2	108.8	114.6	(NA)	59.3	95.7	107.6	111.8
Candy and other confectionery	54.3	81.5	93.8	103.0	98.6	51.2	78.3	93.5	101.7	97.8
Canning and preserving	56.1	79.9	95.6	(NA)	(NA)	58.6	80.5	96.8	(NA)	(NA)
Cement, hydraulic	43.7	68.4	94.5	107.0	109.0	47.3	71.5	94.8	107.8	109.4
Concrete products	56.9	74.4	93.7	109.1	(NA)	58.9	75.9	92.8	109.2	(NA)
Corrugated, solid fiber boxes	(NA)	79.0	94.1	102.1	100.7	(NA)	79.7	94.4	102.4	101.7
Flour and other grain-mill prod	53.0	78.6	99.0	104.1	100.4	54.7	78.0	99.6	104.8	109.3
Footwear	80.6	98.5	101.6	103.8	99.1	80.6	97.6	101.7	103.8	98.5
Glass containers	79.2	83.8	98.8	106.3	113.3	80.3	83.6	98.7	105.0	110.6
Gray iron foundries	(NA)	87.8	102.4	106.4	107.9	(NA)	86.1	103.1	106.3	108.0
Hosiery	39.3	58.1	79.4	92.3	106.7	40.1	58.6	80.0	92.2	105.9
Malt liquors	48.5	67.9	91.6	105.4	111.2	48.0	68.0	91.7	106.0	111.6
Man-made fibers	(NA)	71.7	91.6	109.5	109.3	(NA)	71.8	92.4	113.4	110.9
Motor vehicle and equipment [3]	(NA)	78.5	96.1	107.1	104.3	(NA)	79.6	99.1	109.3	106.5
Paper, paperboard, pulp mills	53.0	73.8	96.3	105.7	111.2	54.7	74.7	96.8	106.1	109.5
Petroleum refining	36.9	62.6	90.3	104.7	116.4	39.4	63.4	90.8	104.1	111.7
Primary aluminum	40.8	82.1	96.3	96.6	99.8	48.3	81.0	96.8	98.4	100.3
Primary copper, lead, and zinc	78.7	96.4	114.4	113.5	116.9	83.2	99.7	121.3	118.6	128.6
Radio and TV receiving sets	(NA)	89.4	93.0	117.6	116.1	(NA)	96.0	94.3	116.6	111.9
Steel [3]	72.5	82.3	98.7	104.6	104.9	78.3	82.3	101.1	104.2	104.8
Sugar	49.6	73.1	95.3	102.5	98.0	51.6	73.4	96.2	103.2	97.0
Tires and inner tubes	50.9	68.6	95.1	105.4	103.6	54.4	69.9	96.6	107.2	106.7
Tobacco products	57.5	81.6	97.6	101.8	104.8	59.6	82.8	98.1	101.6	103.6
Mining: [1]										
Coal mining [3]	37.5	68.0	93.0	105.3	106.6	(NA)	(NA)	(NA)	(NA)	(NA)
Copper (crude ore)	58.9	87.3	104.4	109.8	116.4	(NA)	(NA)	(NA)	(NA)	(NA)
Copper (recoverable metal)	76.6	94.2	106.7	103.3	107.1	(NA)	(NA)	(NA)	(NA)	(NA)
Iron (crude ore)	44.5	68.0	96.5	109.9	117.0	(NA)	(NA)	(NA)	(NA)	(NA)
Iron (usable ore)	76.0	85.1	103.3	108.0	108.8	(NA)	(NA)	(NA)	(NA)	(NA)
Transportation and utilities:										
Air transportation	(NA)	(NA)	(NA)	(NA)	(NA)	[4] 27.1	[4] 52.3	[4] 83.7	[4] 104.3	[4] 107.2
Gas and electric utilities	[4] 28.9	[4] 63.9	[4] 88.7	[4] 107.1	[4] 114.6	31.3	65.6	89.4	107.2	114.1
RR transportation (car miles) [5]	49.7	75.1	93.0	101.9	103.3	31.5	75.5	92.9	101.9	103.4
RR trans. (revenue traffic) [5]	40.4	63.2	91.0	104.4	108.8	42.0	63.6	90.8	104.4	108.9

NA Not available. [1] Man-hours worked, except as noted. [2] Man-hours paid. [3] Includes anthracite and bituminous coal. [4] Refers to output per employee. [5] Refers to output per nonsupervisory worker man-hour. [6] Class I line-haul railroads and switching and terminal companies.

Source: Dept. of Labor, Bureau of Labor Statistics; *Indexes of Output Per Man-hour for Selected Industries 1939 and 1947–69* (Bulletin No. 1680).

Table 2.7 *Comparison of Levels of Output, about 1960*

Country	Real Output per Worker	Real Output per Capita
Argentina	35*	27
Australia	62*	72
Brazil	21*	20
Canada	92*	74
Chile	29*	24
Denmark	62	62
France	52	52
Germany (West)	46	51
Great Britain	50	58
Ireland	36	39
Italy	37	29
Japan	31	27
New Zealand	88*	90
Soviet Union	25*	25
Sweden	57*	65
United States	100	100

*About 1950, rather than 1960.

Before becoming alarmed, one should put these figures in the right perspective. In 1971, for the first time in this century, we imported more than we exported—by $2.9 billion. The net result, at least in theory, was to displace $2.9 billion worth of domestic goods with foreign imports, and to reduce total U.S. output and employment accordingly.

Total U.S. output in 1971 was well over $600 billion. The possible loss in output attributable to the $2.9 billion trading gap was, consequently, 0.5% of the total, and the presumable loss in employment was about the same (i.e., less than one-half of one percent).

The facts about the "outflow of U.S. capital and technology" are equally plain. In 1971, the capital outflow, the additional investment made by U.S. companies in foreign operations, amounted to $4.5 billion. But the capital inflow, the return on previous investment, reached $7.3 billion. This left a positive balance of $2.8 billion.

Similarly, the previous export of U.S. technology produced a cash inflow, in the form of royalties and fees, that amounted to $2.0 billion in 1971.[5]

The real motivation for greater and greater productivity in the United States is that other countries do not have to equal our outputs per man-hour in order to dispossess us in the market place. This is primarily due to the considerable differences in the wage rates paid. Table 2.8 illustrates these differences between selected countries in 1972.

Table 2.8 *Average Wages for Industrial Workers*

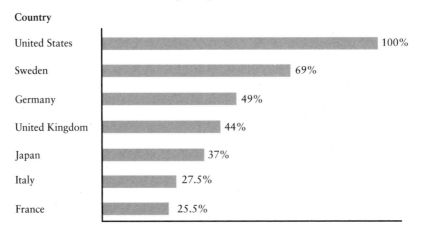

Country

United States — 100%
Sweden — 69%
Germany — 49%
United Kingdom — 44%
Japan — 37%
Italy — 27.5%
France — 25.5%

The wages shown in Table 2.8 are averages for base wages of industrial workers excluding all fringe benefits. The fringe benefits amount to about 20% of the base wage in the United States and about 70% of the base wage in West German.[6]

Comparing Table 2.7 and Table 2.8, it may easily be concluded that many of the competing countries are able to usurp the U.S. markets, even though their productivity has not yet reached the output per man-hour that we enjoy in the United States. Last year, for example, the electronics industry asked for Federal export aid to ward off the onslaught of lowered prices from Japan and Germany.[7] In the automotive industry foreign competition has also made itself felt by moving from 6.2 percent of the market in 1964 to over 16 percent in 1971.[8] Clearly, then, there is a need to find ways and means to improve our productivity.

This year, the average U.S. factory worker will have working for and with him or her the electrical power equivalent of 42 helpers; 50 years ago, he or she had but two.[9] The next 50 years should be equally dramatic. Automation and robotry will play a key role in the improvement of this productivity and in the meeting of this challenge.

Ways to Increase Productivity

Improvements in productivity can be accomplished in a number of ways:

- Workers working harder.
- More capital to provide more productive labor-saving devices, machines, processes, or planning techniques.
- Workers learning new skills to improve the value of their output.
- Workers better organized into production teams, with less time lost in nonproductive effort.
- Fuller utilization of plants to optimum capacity.
- Greater incentives offered by government to industry to invest in more productive equipment and training for workers.

The United States has long been a leader in automated manufacturing and technology. The gap between the United States and other countries is rapidly closing, however, as during the last decade, U.S. industry invested only 13 percent of its gross national product, whereas Japan invested 27 percent and Germany invested 20 percent in new and improved tools and machinery.[10]

References

1. Economic Productivity. In: *Encyclopedia Britannica*. 1972, VII:930.
2. Attacking the Productivity Problem. *Manufacturing Engineering and Management*. Vol. 70, No. 1:27.
3. Kornfeld JP, Magad EL. Robotry and Automation—Key to International Competition. *Proceedings of Second International Symposium on Industrial Robots*. Chicago: IIT Research Institute, May 1972:35-36.
4. Attacking the Productivity Problem. *Manufacturing Engineering and Management*. Vol. 70, No. 1:26.
5. Emery JR. Do We Need a Chinese Wall Around America? *Electronics*. August 14, 1972; Vol. 45, No. 17:172–173.
6. Warnecke HJ, Schraft RD. State of the Art and Experiences with Industrial Robots in German Industries. *Proceedings on Second International Symposium on Industrial Robots*. May 1972:22.
7. Connolly R. Industry Asks Federal Export Aid. *Electronics*. July 17, 1972:53.
8. *Ward's 1972 Automotive Yearbook*. Detroit: Publisher, 1972:63.
9. Kornfeld JP, Magad EL. Robotry and Automation—Key to International Competition. *Proceedings of Second International Symposium on Industrial Robots*. Chicago: IIT Research Institute, May 1972:31.

10. Kornfeld JP, Magad EL. Robotry and Automation—Key to International Competition. *Proceedings of Second International Symposium on Industrial Robots*. Chicago: IIT Research Institute, May 1972:35–36.

14

3

Comparisons between Robots and Fixed Automation

Perhaps the best way to make comparisons between the application of industrial robots and fixed automation is by illustrations from actual practice.

Case 1—Assembly of Liquid Level Monovial[1]

This case deals with the assembly of liquid level monovials, which are subsequently used in carpenters' levels and other similar tools. Figure 3.1 illustrates the basic components of the assembly and also shows three completed assemblies. The empty plastic vials have one end closed and have been molded with a concave ellipsoidal surface on the inside so that the bubble in the liquid will seek the highest portion of the cavity. Once the vials have been partially filled with a measured amount of fluid, a plastic cap is ultrasonically welded to the vial, sealing in the liquid. This completes the assembly.

Early manufacture entailed the manual handling of these vials to do the orienting, placing in bench fixtures, initiating the filling cycle, removing from the filling station and transferring to the ultrasonic welding station, orienting and loading of plastic cap, initiating weld cycle, unloading of completed assemblies, and loading into waiting containers.

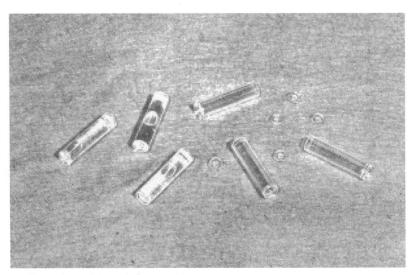

■ **Figure 3.1** *Monovial assembly components.*

One operator could manage to complete a monovial assembly every 15 seconds. This rate does not take into account two 15-minute rest periods, lunch periods. or other times the operator may be away from the workstation. Yearly operation labor costs are approximately $10,000.

Option 1

Two robots, with orienting vibratory feeders for the two plastic parts, could be programmed such that one handled the filling operation and the other took care of the welding station. Capital investment costs were estimated to be about $30,000.

Production rate was improved with these robots to about one monovial every six seconds.

Option 2

Fixed automation was proposed to handle the assembly on a six-station, rotary index table machine, which could handle the monovials at a rate of 32 assemblies per minute or 1.87 seconds for each assembly. This rate was paced primarily by the filling device, which filled the vials with a measured amount of fluid. The cost for this fixed automation approach was $50,000. This machine is illustrated in Figures 3.2, 3.3, and 3.4.

■ **Figure 3.2** *Automatic assembly machine.*

■ **Figure 3.3** *Details of automatic assembly machine.*

■ **Figure 3.4**   *Rear view of automatic assembly machine.*

Manual Economic Comparisons

Production requirements demanded a yearly total of two million assemblies. To achieve this, six operators were used, for a total labor cost of $60,000. With material costs of approximately $0.015 per assembly, this would give a factory cost of about $0.015 + $0.030 or $0.045 per assembly. Maintenance and capitalization of hand fixtures amounted to another $0.005, to give a total factory cost of $0.05 per assembly.

Option 1—Two Robots

Because the production rate with two robots is 10 units per minute, and they can be run for eight hours a shift instead of seven, they will turn out 2,400,000 assemblies on a two-shift basis. Assuming capitalization of the entire cost over five years, the per assembly cost amounts to about $0.032, or a savings of about $43,200 per year.

Option 2—Fixed Automation

This option turns out 32 assemblies per minute or 1,920 per hour. On a single-shift, five-day week, at 100-percent efficiency, this would produce about 3,840,000 assemblies. Capitalizing the equipment over its 10-year life would produce manufacturing costs of $.018 per assembly, or a total savings of $122,880 per year.

This case was selected because it is somewhat typical of those high-volume applications where the parts to be assembled are of uniform high quality and where frequent changes for different sizes or different models are not required. In general, where short-run production is needed, the industrial robot can become more economically justified. It is interesting to note that on the index machine described previously, a simple pick and place vacuum pickup arm has been used to transfer the tiny plastic cap from the feed track to the top of the liquid-filled vial. This robot-like transfer can be done with simple cam timing and actuation because it does not require the flexibility of rapid program changes.

It is important also to note that, frequently, when the cost of robots and fixed automation is compared on a total yearly production rate basis, the cost of the fixed automation can often be less than the cost of the number of robots needed to provide the same rate.

19

Case 2—Tending Die-Casting Machines

This case is a good example in which the industrial robot may be the best solution to automating the production operation. The production requirement is such that two die-casting machines are required on a two-shift basis to handle the demand. Normally, one operator would be used per machine plus another operator on a trim press to remove the sprue and flashing. Manual operators with fringe benefits present a labor cost of $11,000 per year. Obsolescence of parts require a model changeover once every year.

Option 1—General-Purpose Robot[2]

One single robot, centrally positioned between the two die-casting machines and the trim press, can be programmed to handle all three operations. Figure 3.5 illustrates how this would be done. After each die-cast operation is completed, the robot would remove the casting from the machine and place it into the trim press for trimming. The trimmed part is then removed from the trim press by the robot and deposited in the gondola container next to the press. The next part of the program would repeat the operation with the second die-casting machine.

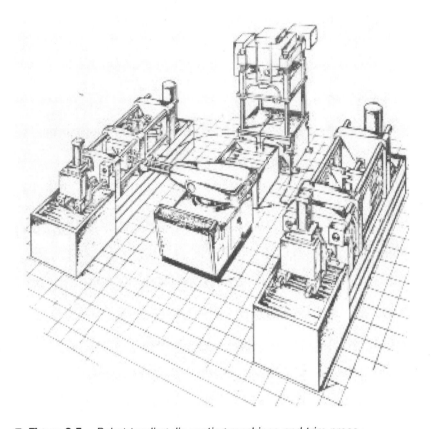

■ **Figure 3.5** *Robot tending die-casting machines and trim press.*

Economic justification became easy for this application, as the yearly savings in labor alone amounted to 6 × $11,000 = $66,000 on a two-shift basis.

$$\text{Payback} = \frac{\$25,000}{\$66,000 - \$3,000} = .4 \text{ year}$$
$$\text{(yrly. maint.)}$$

Return on Investment

$$\frac{\text{Cash Flow}}{\text{Investment}} = \frac{\text{Yrly. Wages-Maint.} - \text{Yrly. Depreciation}}{\text{Investment}} =$$

$$\frac{\$63,000 - \$5,000}{\$25,000} = 230\%$$

Present Value of Future Earnings

(Earnings) (Present Value Factor) = $63,000 × 3.256 = **$205,000**

Option 2—Fixed Automation

Conventional press unloaders and handling devices for this application would have to be redesigned every year, due to changes in the shape and size of the part being manufactured. Also, due to the nature of the die-casting process, three machines would be required plus transfer conveyors and orienting devices. System costs for this type of system are around $60,000. Yearly retooling for each new part is estimated to be $12,000. Maintenance costs are comparable to the robot at $3,000 per year.

$$\text{Payback} = \frac{\$60,000}{\$66,000 - \$3,000} = .95 \text{ year}$$

Return on Investment

$$\frac{\text{Cash Flow}}{\text{Investment}} = \frac{\$66,000 - \$3,000 - \$12,000 - \$12,000}{\$60,000} = 65\%$$

Present Value of Future Earnings

($66,000 − $3,000 − $12,000) (3.256) = **$166,000**

References

1. ITT Industrial and Automation Systems Division Job Files, February 1973.
2. Lindbom TH. Robot's Capabilities and Justification. *Manufacturing Engineering & Management.* July 1972:18–19.

4

Where

The general domain of a robot has been depicted by Lindbom to be somewhere between the manual operator and special automation (see Figure 4.1).[1] It should be understood that the boundaries listed in Figure 4.1 are overlapping in many cases and that, in the final analysis, each application will have to be judged on its own merits. Safety and health reasons or an adverse environment may outweigh the economics of a situation and dictate that a robot be used in place of a manual operation.

Human versus Robot Motion Times

Professor Hangawa of the Institute for Research in Productivity, Waseda University, Tokyo, Japan, has done considerable work in the analysis of human movements and in comparing human capabilities with robots. Results of his work are summarized in Figure 4.2, and in the following quoted conclusions:[2]

1. For carrying light weights (under 1.13 kg) humans are faster than robots. This is especially marked when the area of motion is within the best working area for human beings (within about a 40 cm arc from the operator's shoulders).

2. The difference in motion time between humans and robots becomes almost negligible when the motion distance exceeds 50 cm for the lightweight range. Over 70 cm, humans require more time.

3. For heavy weights (10.5 kg) human motion time is longer in all ranges of distance, and the time increases in proportion to the increase in distance.

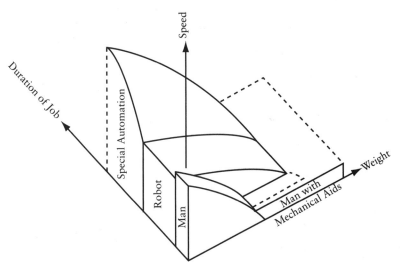

■ **Figure 4.1** *Domains and capabilities of equipment.*

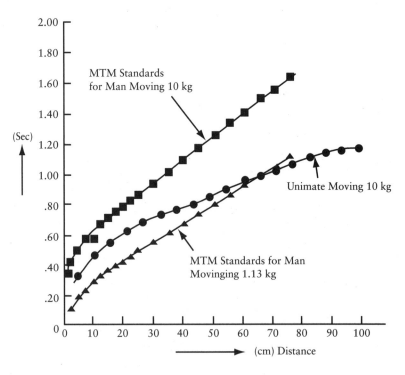

■ **Figure 4.2** *Comparison of human and robot motion time.*

4. Robot motion times show no advantage when the weight is within 10 kg.

5. The most favorable area for robot application is that with greater weight or resistance and distance exceeding the ideal human working area.[3]

In the writer's experience, the province of special automation extends both upward and to the right in Figure 4.1. Where greater speed is required, and where weight exceeds the design limits of the robot, automation designed specifically for the job at hand has been the best solution.

Problem Definition

A most important step in robot application lies in the problem definition stage. One may use Lindbom's job profile technique or Hangawa's matrix approach,[4] but a clear, concise statement of the system specifications and requirements is in order. Some of the important parameters to determine are:

1. **Speed.** Production rate in parts per hour, assemblies per hour, or cycles per hour.

2. **Weight to be handled.** Pounds or kilograms.

3. **Accuracy required of the operation.** This often is a key item in efficient assembly.

4. **Number of axes required.** Often the part may be redesigned to simplify and reduce the axes needed.

5. **Number of programs.** A large number of differing products to be assembled might make a library of programs advisable.

6. **Program length.** The steps to make a complete movement cycle of the robot must be within the memory capability of the programmer in the machine.

7. **Programmability required.** This characteristic has to do with the ease with which the robot may be directed and redirected to do its job. An easily programmed robot is programmed by leading its hand through the desired pattern of motion, or choreographic programming. This may be accomplished with either the continuous-path type of program or the step type, where the end points of each motion are recorded, and playback selects the most direct path between these points.

8. **Reach and volume through which the arm may move.** Most manufacturers will list the maximum reach and describe the effective volume that may be utilized by the arm. These are typically spheres or parts of spheres, cylinders, or rectangular boxes.

9. **Type of hand or hand fixture required.** Figure 4.3 illustrates some of the various hands and grippers that have been used successfully.

10. **Environmental factors.** Frequently these factors play a dominant role in causing management to seek a robot for a particular job. High nuclear radiation levels, noise levels, heat, dirt, hazardous atmospheres, safety hazards such as explosives, or bomb manufacture may easily outweigh the economic factors in robot selection.

11. **Initial cost.** Although prices of robots will gradually be reduced as greater volume is achieved, this is a prime factor in economic justification.

12. **Maintenance and operating costs.** The total economic justification equation would not be complete without this very important factor. It probably is one of the most elusive to establish. Good data are usually known on the amounts of electrical power, air, hydraulic, and lubrication requirements, and the wear and spare parts consumables. But the big question mark is the potential down time and the maintenance time required, as this is usually a reflection on the system design, rather than the mechanical design of the robot. In a private interview with Mr. "Andy" Anderson, general manager of GM's highly automated Lordstown, Ohio, Vega plant, he stated that: "If he had his druthers, he would probably go back to manually welding the Vega bodies instead of using the present 28 industrial robots."[5] This is primarily due to the current high maintenance costs and the downtime caused by nonadaptability of the robots to changing production conditions. Maintenance costs from robot manufacturers have been hard to come by, as these sometimes cast an unfavorable light on the robot application. Likewise, robot users are reluctant to divulge this type of data for the same reason or because it may reveal proprietary information to competition. Normal maintenance costs, for a robot properly applied, average from 10 to 15 percent of the initial capital investment.

Lindbom has perceived that "a rule of thumb to follow is that there should be a potential for at least three to five industrial robots in a plant before one is purchased. The reason? The job planning techniques used, the presence of someone on every shift who can start and stop the robot, the periodic maintenance and service, are overhead factors that are minor when there are multiple robots in a plant. However, this overhead becomes large relative to savings resulting when only one robot is used."[6]

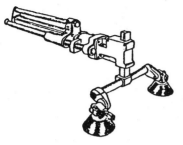

Different hands without wrist movements
(ISI Man. Co.)

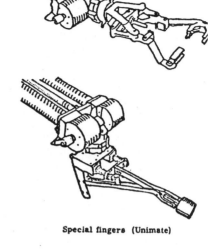

Special fingers (Unimate)

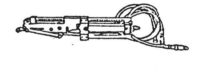

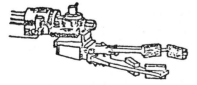

Hands for a pure mechanical pick and place
unit (Orii)

27

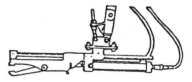

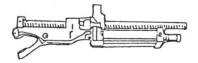

■ **Figure 4.3** *Views of typical robot hands and gripping devices (taken from Robot Survey).*

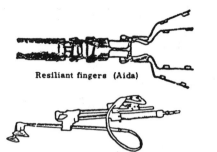

Resiliant fingers (Aida)

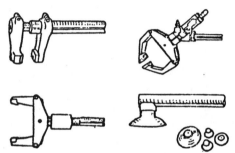

Three vacuum lifting beams for parts with a
curved surface. (Tsubaki)

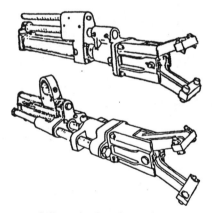

Three different clamping devices and vacuum
pads. (Auto-place)

Different hands without wrist movements
(ISI Man. Co.)

■ **Figure 4.3** *Views of typical robot hands and gripping devices (taken from Robot
Survey) (continued).*

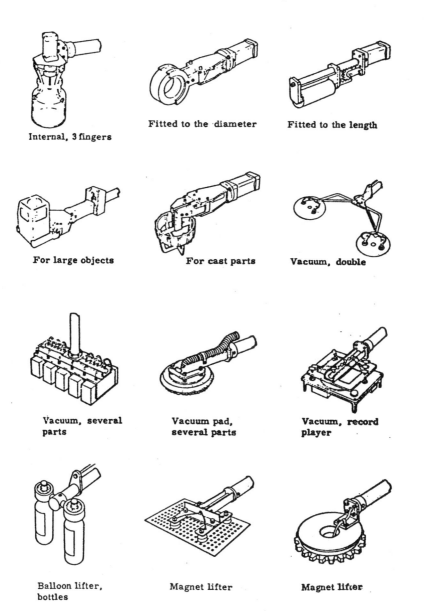

Internal, 3 fingers

Fitted to the diameter

Fitted to the length

For large objects

For cast parts

Vacuum, double

Vacuum, several parts

Vacuum pad, several parts

Vacuum, record player

Balloon lifter, bottles

Magnet lifter

Magnet lifter

■ **Figure 4.3** *Views of typical robot hands and gripping devices (taken from Robot Survey) (continued).*

Case Studies

Following are a number of case histories from the files of some of the foremost suppliers of robotic systems. They have been selected to give a cross-section of those systems supplied in the 2000 to 2004 time period.

How Robots Are Used in the Composites Industry

Increasing costs and environmental regulations have forced composites manufacturers to consider robotic automation for the application of fiberglass chop and gel. In a recent process examination, robots provided several key benefits over manual application methods (see Glasland case study below).

Material Savings

Variations in the amount of material used for manual application of resin or fiberglass lead to excessive expenses through warranty costs, part re-work, production delays, part fit, and performance problems.

FANUC Robotics' P-series robots and controllers are designed specifically for industrial coating applications such as fiber reinforced plastic processes. Robotic application provides consistent material delivery control, which reduces material variation and increases part production consistency. As a result, material costs and lay-up time are significantly reduced.

Labor Savings

Manual fiber chop applications can lead to repetitive motion injuries and prolonged exposure to resin fumes, and fiber chop can potentially cause other physical complications. In addition, strict EPA regulations to employee exposure and to open molding materials styrene emissions have increased.

Product Quality

Inconsistent application of gel coat results in the reduction of product quality and performance. Gel coat applied too thickly can cause product cracking; if it is applied too thinly, the gel coat may not cure properly, or bleed-through of the fiber reinforcement may occur. Automation ensures that materials are applied consistently, which helps manufacturers improve quality and reduce warranty costs.

Automated Solutions

FANUC Robotics understands the unique application challenges faced by the composites industry. Our engineering staff specialized in the development of turnkey solutions that address critical labor, quality, and cost savings issues.

The Process

Before partnering with an automation company, it must be determined that the supplier has the process experience and application expertise to meet the demands of the fiberglass market. The following list describes many of the process benefits that a qualified automation supplier will be able to provide to composites manufacturers.

- Extensive experience within the fiberglass open mold forming market.
- Reduced material overspray via accurate trigger timing and precise robot motion.
- Resin application flow and glass feed rates controlled to a tight, fixed set point.
- Significant reduction in styrene emissions to help increase EPA compliance.
- Real-time production monitoring.
- Improvements in the visual appearance of gel coat and fiberglass chop thickness uniformity.
- Reductions in warranty costs, part re-work, production delays, part fit, and performance problems.

System Specifications

FANUC Robotics offers the following products and support services to composites manufacturers:

- P-Series robots offer the flexibility and reach required for industrial painting.
- Teach pendant provides full functionality for setup, troubleshooting, and path teaching.
- Fully integrated flow meter system monitors and controls material application.
- Glass detection and breakage feature mounted to gun; ensures proper material application.
- Advanced controller allows gun trigger accuracy of 4 milliseconds.
- Complete lab facilities for system design, build, and customer demonstrations.

Optional System Features

- **Paintworks stand-alone.** Optional Windows-based graphical off-line path and program modification package. Paths created by Paintworks may be directly loaded to the robot.
- **Gel coat color change.** Allows on-the-fly color and gel coat type change. The robot handles line clean out and purge.
- **Closed-loop material delivery.** Accuchop closed-loop material delivery system monitors and maintains consistent fiber chop and gel coat application and makes required adjustments based on robot feedback.
- **Mirror imaging.** Requires only one path to be taught for symmetric parts; mirrors path for opposing robot.
- **Networking communications.** Wide range of networking communications available: Remote I/O, Host Communications, and Distributed I/O. Optional Ethernet allows path data communications through FANUC File Services software.

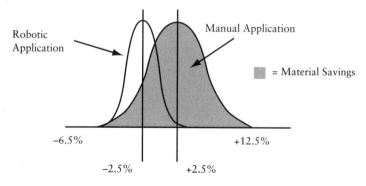

- **Figure 4.4** *Typical example of material savings by weight.*

Customer Case Study

Robots Improve Productivity at Glasland Industries

Glasland Industries, Inc., a Silver Lake, Indiana, manufacturer of fiberglass products, recently converted its chop and gel coating area from manual operations to robotics. The conversion to robotics took place after Glasland was awarded a large contract by an egg producer to manufacture the fiberglass housings of ventilation fans used to control the climate in hen houses.

The contract's aggressive production plans meant that Glasland would need to hire additional skilled chop and gel coat operators, but none were readily available in the local workforce. The lack of skilled labor and high production rate requirements prompted Glasland's management to evaluate robotic automation.

Glasland selected FANUC Robotics North America, Inc. as the robotics supplier. According to Doyle Heckaman, president of Glasland, the company selected FANUC Robotics based on its reputation for reliable products and superior customer support.

Glasland's system uses two robotic paint work cells supplied by FANUC Robotics' paint operations in Toledo, Ohio. Each cell includes a P-145/RJ-3 paint finishing robot with PaintTool application software and safety equipment including safety fences with interlocked gates and curtains, and servo disconnect switches. FANUC Robotics also supplied an explosion-proof manual input station, which is used by technicians to set program requirements and to start the robots, and solenoid valves that turn the spray guns on and off. Glasland supplied the chop and gel coat application equipment, including the spray guns, chop motors, and material supply pumping systems.

While Glasland chose to do its own system integration, FANUC Robotics was on-site for installation supervision, system start-up, and debug activities. In addition, a process engineer from FANUC Robotics did the initial robot programming and a trainer provided on-site programming and operations training to Glasland's personnel.

"We assigned our best operators to program the motions of the robots," said Heckaman. The repeatability and consistency of the robots allows the skilled operators to tackle other jobs at the company. Glasland has technicians who prepare the molds and position them in front of the robots. They also start the robot cycle and monitor their operation and the chop and gel coat application equipment. While the robot applies the chop and gel coat, the technicians can prepare the next molds for production operations. "So far the robots have maintained nearly 100 percent uptime," added Heckaman.

Since the system was installed in July 1999, Glasland and its employees have been happy with the robots' performance. "The fiberglass industry has been using the same equipment for 30 to 40 years," remarked Heckaman. "The technology has not really improved until now, but with robots, we're producing better quality parts, reducing costs, and meeting our customer's production schedules on time."

Improved accuracy has been the major factor in part quality improvements and reduced costs. In the past, even the most highly skilled chop and gel operators became fatigued by the repetitive motions required by the application. With fatigue comes the likelihood of overspray and the extra costs associated with material wastes.

For chop and gel applications, quality is assessed by the uniformity of material weight. The P-145 robots, capable of achieving ± 0.5 mm repeatability, spray materials within a 1 percent variation of weight. In addition, Glasland has realized a 30 to 35 percent savings on gel coat materials. "They also never tire and rarely make a mistake," added Heckaman.

"No comparison," was Heckaman's first comment when asked how he would compare the robotic system to the manual system. "Manual operations were slow and very physically demanding. We were also concerned by the high amount of human exposure to potentially harmful emissions and fiber chop. The robots eliminate these issues."

How the System Works

The production of one fan housing requires two gel coat applications and one chop spray application. The following describes how the system works step-by-step.

- A technician prepares a mold for the initial gel coat application and positions the mold in front of the gel coat robot. The technician sets the safety interlocks, checks to ensure that no personnel are in the robot work area, closes any gates or curtains, and energizes a servo lock-out switch. After the robot cycle is complete, the technician moves the mold to an area where the first gel coat application can cure.

- After the first gel coat application is cured, the technician positions the mold in front of the chop robot for the chop application. The safety procedures are adhered to before the robot begins the fiber chop spray. Upon completion of the chop spray cycle, the technician positions the mold in the chop roll-out area.

- When the roll-out is complete, the part is moved to an area for drying and then to a grinding booth. After the surface of the part is ground and smooth, it is moved back to the gel coat area.

- The technician positions the mold in front of the gel coat robot for a second gel coat application. Safety precautions are again adhered to, and the robot applies the final layer of gel coat. Upon completion of this cycle, the technician moves the mold to the curing area.

- After curing, the part is removed from the mold and transferred to an area where other operations are performed to complete production. Finished fan housings are loaded onto a truck for shipment to the customer.

Production speed has been favorably rated by Glasland. The robots are able to accomplish a day's worth of work in about half the time it takes a fully manual operation. In fact, the robots have not yet been required to run at their full capacity.

"The system has also given us more flexibility with our labor," added Heckaman. "We've been able to improve working conditions and we no longer have to worry about absenteeism in the spray area—the robots are here every day."

Overall, the robotic system has allowed Glasland to have a competitive advantage in the composites industry, where robots are just now beginning to be used. With the increase in production, material savings, and reduction in labor costs, payback is expected to be just two years.

Most important, Glasland is able to provide its customers with products of the utmost quality and appearance. While production does not require additional robots at this time, would Glasland purchase more robots to meet increased demand? "Yes, we definitely would," said Heckaman.

Glasland Industries was founded in 1972, by Doyle Heckaman. Mr. Heckaman's business began as a small home operation, which provided tooling for the RV and boat industries. In 1978, Glasland was incorporated and the company grew at a rapid pace. In 1987, Glasland moved to its current location.

Glasland offers a diverse range of products. In addition to fan housings and van conversions, the company designs and manufactures Splendor Boats. The Splendor Boat line includes full-planing, catamaran hull boats in lengths of 20 feet to 28 feet, and a 14-foot fishing boat.

■ **Figure 4.5** *FANUC Robotics' six-axis P-145 robot consistently applies fiberglass chop spray to fan housings manufactured at Glasland Industries.*

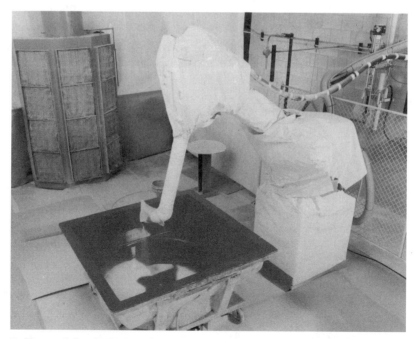

■ **Figure 4.6** *Robotic automation has helped Glasland realize a significant savings on gel coat materials.*

Why Buy Robotic Vision from a Robot Supplier?

When evaluating a robot application that includes a machine vision component, one critical decision is where to purchase the machine vision system. Should the system be purchased from a robot supplier, or is it better to purchase from a vendor who sells general-purpose machine vision? The following issues are presented to help end-users and integrators make the best decision for their application.

38

How Much Spplication Time Is Required to Get a Robotic Vision System Working?

Products available from robot suppliers are specifically designed for robotic applications with features tuned for robotics. The vision portion of robotic applications can be set up using a graphical interface with no programming required. Typically, vision is set up within a half day or less.

A general-purpose machine vision system will have a broader range of options, but few specifically related to robot vision. Setting up robotic vision with a general-purpose vision system requires extra time to sort out what features should be used for the application, and extra time for the significant amount of programming required. One general-purpose machine vision supplier proudly claimed that a robotic vision application only took two months to develop. The same application using visLOC, FANUC Robotics' vision product, would have taken one-half day.

Another example of development time required when using a general-purpose machine vision system is found in the article "Machine vision speeds robot productivity," (Vision Systems Design, October 2001, page 70). A featured robotic vision application took a systems integrator approximately five months to develop.

How Is Calibration Established between the Robot and the Vision System?

Products available from robot suppliers have built-in, easy-to-use robotic vision calibration based on either calibration grids or motion-based calibration. The calibration supplied with general-purpose vision systems tends to be clumsy at best. One supplier required manual input of robot points within a separate graphical panel for each point in the calibration. Not only is this awkward, but it is also prone to error.

What if a Robotic Vision Application Has Some Subtle Complexity?

The easiest application is a fixed camera reporting the horizontal or vertical position of an object that is not held by the robot, and using this information to offset one robot position. An example of a more complex application is shown in Figure 4.7. For this application, as the robot moves the automotive floor pan under the camera, an image is acquired at both ends of the part. From the two images, the calculated robotic offset allows accurate placement of the floor pan into a rack. Many applications require or benefit from more complex setups that may include:

- Rotating an angled gripper.
- Offsetting an entire robot path.
- Transforming to alternate robot frames (used or tool frames).
- Putting a part or camera on the robot.
- Requiring multiple camera views to image a part.
- Adding 3-D input through stereo or other imaging techniques.

With the type of complexities listed above, applications are far easier to set up with a robotic vision system that has the required built-in features. These features are available when purchasing vision from a robot supplier. Extensive programming is required to compensate for missing features when using a general-purpose vision system for robotic applications.

■ **Figure 4.7** *Imaging an automotive floor pan held by a robot. The camera is located in the black protective housing at the top of the figure.*

How Easy Is it to Program the Robot for a Robotic Vision Application?

When using general-purpose vision for robotic applications, there is no built-in support for programming a robot. On the other hand, with a robotic vision system such as visLOC, there is significant built-in support. The snippet robot program shown on the teach pendent (Figure 4.8) illustrates this. The "Snap&Find" command identifies the vision process used and does all the required hand shaking between the robot and the vision system. Registers used by robotic vision (position, control, and measurement registers) are automatically defined. All of the registers are updated, both data and label, by the "Snap&Find" command and other vision commands.

■ **Figure 4.8**   *Robotic vision program on a teach pendant.*

What Happens if Something Goes Wrong?

When there is a problem, users must determine if the problem is caused by the robot application or the vision application. If the vision is purchased from a robot company, it is easy to decide where to go for help. If a general-purpose vision system is used, there is no easy way to determine if the problem is with the robot, the vision system, or some combination of robot and vision. In addition, if the vision was purchased from a robot supplier, it is likely that the supplier has experience with similar applications. Thus, the robot supplier, based on previous applications, is able to supply a "cookbook" approach to the problem and find a solution more quickly.

Deciding on whether to use robotic vision purchased from a robot company or to use a general-purpose vision system comes down to asking the following question. "Are the broader range of options available in the general-purpose system worth the extra time and cost required to make use of them for a robotic vision application?" As a general rule, the more experience system integrators have with vision, the more appreciation they will have for the value of a "ready-to-use" robotic vision system. The on-time delivery of a reliable, maintainable, and cost-effective vision solution enables integrators to maximize their customers' productivity.

Robotic Sealing System Helps Lamson & Sessions Increase Production Flexibility and Reduce Operating Costs

Lamson & Sessions is a leading producer of thermoplastic conduit, enclosures, wiring outlet boxes, and accessories for the electrical, construction, consumer, power, and communications markets, and large diameter pipe for the waste water market.

43

Based in Cleveland, Ohio, Lamson & Sessions manufactures millions of injection-molded PVC electrical wiring boxes and junction boxes per year. To achieve approval from Underwriters Lab and the Canadian Standards Association, each box requires a gasket seal to prevent water intrusion. For more than 20 years, the company used die-cut foam gaskets with adhesive backing to accomplish the sealing operation. With thirteen size variations of boxes, and an increasing production schedule, the manual application had a number of challenges to be addressed, including:

- Gaskets are sometimes too thin to properly compress inside the boxes. This prevents proper sealing in situations where box lids are slightly warped or "crowned."
- Field installation of gaskets to lids is time consuming and tedious, especially when creases or other irregularities slow down the process of removing the paper backing from the adhesive strip. This problem is compounded in outside installations during cold weather and the need to remove gloves to install gaskets.

- Improperly installed or missing gaskets reduce quality levels and result in customer dissatisfaction.
- Costs of the existing die-cut foam gasket system were very significant and included the high cost of the purchased gasket, inventory/tracking costs for thirteen different models, as well as the labor costs to manually pack the gasket into the junction box (no reference to actual dollar savings, per L&S vice president).

Robotic Alternative

To eliminate the shortcomings of the manual operation, Lamson & Sessions chose to convert the gasket sealing system from cut foam to a urethane "form in place" (FIP) gasket. Chemque, a leading manufacturer of specialty high performance polymers, manufactures the material, which is less costly and offers a better seal than cut foam. Lamson & Sessions also evaluated alternative methods to apply the material and determined that robotic automation would provide the flexibility and precision required to meet production demands.

Lamson & Sessions selected RADCO Industries, a system integrator in Toledo, Ohio, to develop an automated system that would uniformly dispense the FIP gasket and maintain the critical seal on the PVC electrical wiring boxes and junction boxes. The solution required consistency, precision, and flexibility to accommodate the thirteen different gasket sizes.

System Components

The automated dispensing system incorporates a FANUC M-710iB/45 and a FANUC LR Mate 100iB robot. The M-710iB/45 is paired with a two-component meter mix dispensing system with triple servo motor portioning and mix. Some details about the meter mix include: The urethane form-in-place gasket is a two-component catalyzed system delivered in individual 55-gallon drums. These drums are adjacent to the machine and deliver the A or B component on demand with air operated drum pumps. The product is delivered to one of two "day tanks" holding approximately 10 gallons. The A component is the body of the adhesive. Pressurized air is introduced into the tank, and a motor operated paddle mixes the fluid to uniformly distribute the air. This air entrainment will give the final gasket its loft and sponginess. The percentage of air entrainment is monitored and controlled, as well as the mixing paddle motor. The B component, which is the catalyst tank, is equipped with desiccant dryers to eliminate harmful moisture.

On demand, each day, a tank will deliver product through individual electric pumps to the metering system. This consists of an individual servo-driven precision gear pump for each component. The servo pumps allow each component to be delivered in a very precise ration, with an accuracy of approximately 1/100 of a gram. As a reference, a dollar bill weighs 1 gram. Cut a dollar bill into 100 pieces, weigh one piece, and the result equals the delivery ratio accuracy. The components, still handled separately, are brought to a servo-controlled mixing head. The two parts are thoroughly mixed and the catalyzation process begins. This mixing head is maintained at an exact temperature to control the reaction time that is accomplished with a chilled or warmed water manifold. The servo mixer also generates internal pressure in the head. A double solenoid-operated dispense valve opens, as signaled by the FANUC robot, to deliver air entrained catalyzed product onto the electrical box lid. The dispense rate and volume are controlled through dispense servo motor speed (addressable via the robot control), nozzle diameter, and robot speed. All travel and dispense parameters are downloaded and preset when the part number is keyed into the HMI.

The M-710iB/45, equipped with HandlingTool application-specific software, handles the dispensing of the FIP gasket. Articulation and precision are key characteristics of the M-710iB/45, which can handle payloads up to 45 kg. The robot also has an extremely large work envelope without requiring significant floor space.

In addition, the M-710iB/45 offers floor, invert, wall, shelf, and angle mounting to accommodate a variety of installation variations, making it possible for users to have better access to unusual work pieces. The flexibility of the robot's design allows it to flip over and work behind itself. A 70-kg variation, the M-710iB/70, is available for higher payload applications. Both models offer a compact body and six axes of motion.

Once the FIP gasket is dispensed, a five-axis, FANUC LR Mate 100iB robot, equipped with a quick-change vacuum end-of-arm tool, transfers the electrical boxes to an infrared curing oven. The robot's small size and extensive capabilities allow it to easily handle the parts without disturbing the integrity of the dispensed gasket. This automated process helps the customer achieve quality standards that could not be met manually or with hard automation.

The LR Mate 100iB's small arm size offers maximum loading capabilities in the mini-robot category. It offers five axes of motion, and has a payload capacity of 5 kg.

The LR Mate 100*i*B can be used in a variety of applications such as packing, packaging, picking, material handling, machine load/unload, parts cleaning, testing, and sampling. The small size of the robot and its extensive capabilities make it the perfect solution for industrial applications or use in laboratory environments.

RADCO Industries designed and built the curing oven with an automatic stainless steel link belt conveyor. The 20 kW infrared oven cures parts at 160 degrees Fahrenheit and 100 percent humidity. Watlow's 1120 radiant heater panels provide heat control for curing.

RADCO system also specified a Camco three-station indexing dial system to provide a load station, gasket dispense station, and auto unload station.

How the System Works

An operator manually loads parts into the indexing dial system, which has a dedicated master plate to locate each set of tooling. The tooling is made of UHMW polyethylene, a tough material with a chemical resistant surface. As parts are loaded, the system assigns and stores a unique signal for each of the thirteen box sizes. Once tooling location is complete, the table indexes to begin the dispensing operation. A light curtain prevents operator intrusion until each job is complete.

The robot dispenses a bead of urethane foam around each part to form the FIP gasket. Bead rates depend on the mix of the material, the speed of the robot, and the size and features of each part, such as corners, and a "ramped blend," which occur at the start and finish of each gasket. Dispense rates vary from 3 inches per second for the smallest part to 6 inches per second for the larger parts. The foam gaskets typically take 3 minutes to gel.

Upon completion of all of the gaskets, the M-710*i*B/45 stops and the table indexes to the unload position, where the LR Mate 100*i*B transfers parts to the curing oven conveyor. The robot picks from one to six parts, depending on weight and dispense cycle time. The curing oven conveyor has a variable speed drive to maximize throughput regardless of the amount of parts being unloaded.

All three stations function simultaneously: load-dispense-unload. Unload time is always less than dispense time, which makes throughput dependent on dispense speed. Average processing time is 4 seconds per part.

The curing oven reaches the correct temperature and humidity level before dispensing begins. A Simpson temperature indicator, which is tied into the control system, signals the robot to begin dispensing when the oven is ready. Automatic temperature gauging helps streamline the entire process.

Finished products exit the curing oven into completion storage bins, sorted and prepared for shipment.

The Results

Payback for the system is estimated to be just 12 months. RADCO engineers credit preplanning and conceptual design during the quoting stage to the system's success and quick return on investment. RADCO also performed extensive testing of the adhesive and the infrared/moisture curing mechanisms. The curing was a crucial part of the system engineering, as no one had previously taken freshly dispensed foamed urethane gaskets and dropped the parts into bulk storage bins.

The customer has realized several key benefits as a result of the new robotic system including:

- Sizable cost savings, with the elimination of the previous cut foam gaskets.
- A consistent and reliable automated process that significantly increases product quality.
- Flexibility to handle additional parts in the future.

According to George Foos, senior project engineer at Lamson & Sessions, the entire project was a success. "From initial construction to final testing, RADCO Industries accommodated our 'on the fly' changes. The changes proved to be painless and really enhanced the functionality and user friendliness of the system.

"The system is well-conceived and innovative. It's highly durable, as in 'built like a tank.' I'm sure this system will be running production 20 years from now," added Foos.

Lamson & Sessions is a leading domestic producer of thermoplastic conduit, enclosures, wiring outlet boxes, and accessories for the electrical, construction, consumer, power, and communications markets, and large diameter pipe for the waste water market. The company brings its products to the U.S. and Canadian markets through three business segments, divided by channels of distribution and markets served. For more information, contact Lamson & Sessions at www.lamson-sessions.com.

47

RADCO, a member of FANUC Robotics' integrator network, is a 41-year-old turnkey systems house specializing in custom-designed and custom-built equipment for a wide range of applications including assembly, adhesive, and gasket dispense, pressure decay and helium mass spec leak testing, CNC and indexing dial machining systems, and many more. The company has full in-house capability for conceptual, mechanical, and electrical design as well as complete sheet metal and fabrication facilities, panel building, and electrical integration, including multi-axis CNC, plus assembly, test, and onsite start-up. For more information, contact Jim Paul at RADCO Industries at 800-283-0792 or online at www.radcoindustries.com.

FANUC Robotics America, Inc. designs, engineers, and manufactures robots and robotic systems for a wide range of industries and applications. After 20 years of success, FANUC Robotics maintains its position as the leading robotics company in North and South America. A subsidiary of FANUC Ltd. in Japan, the company has facilities in Chicago; Los Angeles; Charlotte, NC; Cincinnati; and Toledo, Ohio; Montreal; Aguascalientes, Mexico; and Sao Paulo, Brazil. FANUC Robotics can be found on the Internet at www.fanucrobotics.com or by calling 800-47-ROBOT.

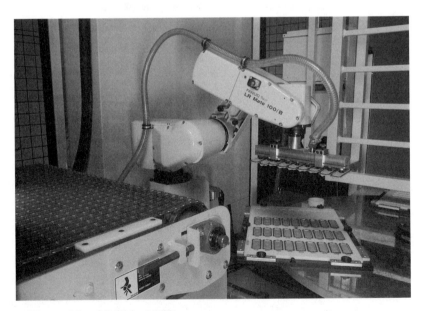

■ **Figure 4.9** *LR Mate 100iB.*

■ **Figure 4.10** *Robot M-710iB/45.*

Survey of Robotic Seam Tracking Systems for Gas Metal Arc Welding

Mark Scherler, FANUC Robotics America, Inc.

A key issue related to robotic arc welding is maintaining consistency in parts placement. A quality weld can only be achieved if the weld seam of each part moves less than one-half the diameter of the weld wire from the programmed weld path. A standard robotic arc welding system does not have the ability to see the changes in the joint location. If unacceptable weld joint variation occurs, an independent method of finding the location of the weld seam must be used to help the robot adjust the position of the weld path. This case describes three common robotic arc welding sensor processes used today.

TAST

For gas metal arc welding (GMAW), a low cost method of tracking a weld seam is through-arc seam tracking (TAST). This method uses the welding arc as a sensor to measure variations in the welding current that are caused by changes in arc length. For example, a change in stick-out is determined by using the inversely proportional relationship between arc length and arc current. By monitoring the arc current feedback, the robot can adjust the torch's vertical position to maintain a constant stick-out. The lateral location of the seam is determined by using the weave function of the robot. As the torch weaves over the seam, the weld current feedback oscillates and indicates the position of the arc.

TAST control systems typically have a large number of variables that are used to optimize tracking performance for a variety of applications, making TAST relatively difficult for operators to program during the integration period. Once the variables are optimized, it is critical that a stable welding process is maintained. Since the weld current is dependent on the weld process, any changes in this process will affect feedback going to the robot and may result in the robot trying to compensate by incorrectly adjusting the torch position.

TAST is an inexpensive method of tracking weld joints. The only hardware requirement is a weld current sensor, which is already included on most welding power supplies.

Laser-based Systems

When material or process conditions create an application that is not feasible for TAST to track a joint, a laser-based sensor can be added to the robot. For sheet metal applications, a laser sensor can track lap joints with material thickness less than 1 mm. To track these joints, the laser sensor must be placed in front of the welding torch to allow the laser to scan across the weld joint. A camera inside the sensor monitors the laser light to determine the location of the weld joint and passes this information on to the robot.

Unlike TAST, the laser sensor does not need to have an arc established to get joint information. The sensor can be used to search for the joint location before starting the weld, allowing the robot to place the wire directly on the joint before the arc start. Once the weld is started, the laser can be used to track the joint and to determine information about the weld joint, such as gap or area. As joint information changes, the robot can adaptively modify weld parameters to match optimal settings for changing conditions.

A laser sensor creates a relatively complex system that can be affected by the everyday rigors of a harsh production environment. Because the sensor package is attached to the weld torch, it can become an obstruction that limits torch access to some joints. To justify the high cost of a laser tracking system, a study should be performed to determine if the laser sensor would provide a significant reduction in weld repair costs.

Touch Sensing

Another method of adjusting a robot path for movement in a weld seam is called *touch sensing*. Touch sensing does not function as a seam tracker, but instead finds the weld seam and adjusts the entire weld path before starting the arc. The robot finds a seam by using the welding electrode or a separate pointer to make electrical contact with the part. The robot performs a search pattern so that it can touch the part to find out how far the seam has shifted and rotated in up to three dimensions. An offset can then be applied to every weld that is on this seam. This method can also be used to determine if there is a gap in the weld joint that requires a change in the weld schedule.

The requirements for touch sensing are very straightforward. Most robotic welding power supplies contain a circuit that can be used for touch sensing, making this a very low cost method. The biggest disadvantage comes from the cycle time increase that is added by the robot to perform the search routines. The weld joint must also have an edge that can be found by the sensor. For lap joints, this requires a top plate thickness of 2 mm.

Conclusion

Each method of sensing presented in this case has its own advantages and limitations. With the right knowledge, users can determine what is needed to gain maximum productivity from their robotic application. Table 4.1 compares each of the sensor methods. It is important to note that a welding sensor will not improve the welding process, but will only maintain correct alignment of the torch to the part. By adding one

Table 4.1 *Sensor Comparison Chart*

	TAST	Laser	Touch Sense
Seam finder	N	Y	Y
Seam tracker	Y	Y	N
Adaptive capabilities	Y	Y	Limited
Joint types	Lap, Fillet, Butt, Ridge	All	Lap, Fillet, Ridge
Material types	All Steels	Nonreflective material	All
Min. lap thickness	2 mm	.8 mm	2 mm
Additional cycle time	none	<1 sec per search	1.5 sec per search
Programming complexity 1–5	4	4	2
Maintenance requirement 1–5	4	3	1
Welding processes	GMAW, Pulsed GMAW, Sub Arc	Most Welding Processes	Most Welding Processes

of these sensor methods to the manufacturing process, additional complexity is introduced to a robotic system. Robotic arc welding can only be simplified if the manufacturing process allows for consistent weld joint placement, thus eliminating the need for sensors.

■ **Figure 4.11**   *FANUC's VAGROBOT Y-16 with seam tracking.*

Well-Dressed
Why the Robotics Industry Is Turning to All-Inclusive Dress Packs

By Mark Eisenmenger, General Manager of LEONI EPS, Troy, Michigan

So what are the well-dressed robots sporting nowadays? For precision and durability, the latest fashion is to dress them with umbilicals, the integrated dress pack that carries all of the power, data, and pneumatic and hydraulic hoses bundled inside a highly flexible, polyurethane jacket.

Over the past four decades, there has been a quantum leap in the performance, reliability, and demands of industrial robots. From a conceptual standpoint, the cabling technology supporting them has lagged. Until recently, the 40,000 or so spot welding and material handling robots in North American auto plants were still being dressed much like first generation robots—with a hodgepodge of customized brackets, hoses, and cables, each with a rather short working life, reflecting the intense conditions to which they are subject.

Enter the umbilical. With close to 4,000 robotic dressout umbilicals installed throughout North America, LEONI EPS is a leader in this niche market. LEONI ESP components are extruded into the umbilical shell to form a linear spring system that absorbs and distributes the stresses and loads throughout the entire module. The umbilical shields the cabling from abrasion, weld sparks, and bending stress. The custom assembly—posts, brackets, boots, and umbilical—is designed to absorb the kinetic energy generated by the robot's operations.

Article for *Robotics World*, November 2003.

Last year, LEONI ESP and robot manufacturer FANUC Robotics North America, Inc. of Rochester Hills, Michigan joined on a plant-retooling project for a major North American automaker. The project involved over 350 robots, with complete dress package, for three application types: spot welding, materials handling, and combinations of both functions. The project set a number of precedents. FANUC and LEONI ESP had worked together in the auto industry, but this was the first customer to require a common dressout that could be attached to any configuration at the end—as opposed to the shoulder, as per conventional consumables—of any arm tooling, from servo guns to material handlers. FANUC and LEONI ESP delivered.

Now, the integrator only has to make final connections when debugging the robot program, saving up to 8 hours per unit during the integration/startup phase.

"It was also the first time a customer required a warranty, which would have to include cables and hoses, normally considered consumables," says Joe Grazzarato, manager, Engineering Body Shop, at FANUC Robotics North America. That's largely unheard of for conventional consumables, but LEONI EPS offers an industry-leading 1-year warranty for cables and hoses because of the robust performance of its umbilicals.

With proper installation, umbilicals can increase cable and hose life tenfold. The umbilical is an extremely precise guidance system. With its flexible brackets, it can be adjusted during robot programming to protect the primary power cable from abrasion, weld sparks, and bending stress. Taking into account all robot trajectories as well as the surrounding environment, each umbilical system can be positioned to avoid contact with cutting edges, rubbing, and other robots.

Another major feature of the specs FANUC and LEONI ESP met was a requirement for standardization. Traditionally, standardization is difficult to achieve within a single application type. There are too many weld gun configurations at an auto plant to allow for standard configurations with conventional consumables. LEONI and FANUC were able to standardize the dress packages for each application type by using a common hardware package. "The flexible modular system allows for easy adjustment and fine-tuning for specific applications, and has allowed us to meet the requirements of our OEM client in a cost-competitive, high-quality environment," says FANUC's Grazzarato. Each dress package uses the same basic mounting and support components, resulting in a shorter parts list and smaller inventories.

With umbilicals offering these benefits and much more, operators with an eye on maximizing the return from multimillion-dollar robotics upgrades are expected to turn to all-inclusive dress packs with increasing frequency.

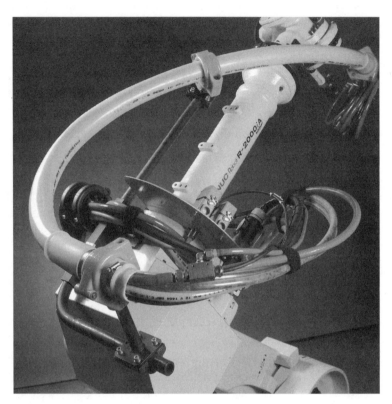

■ **Figure 4.12** *All-inclusive umbilical—holding all cabling and pneumatic and hydraulic hoses needed in robotics (LEONI EPS dresspack).*

■ **Figure 4.13** *FANUC robot with welding gun as end effector (LEONI.FANUC.still).*

Websites for article/readers of *Robotics World*

www.leoni-tailormadecable.com

www.fanucrobotics.com

Successful Robotic Laser Cutting Applications

Great advancements have been made in the application of robots performing laser cutting. Several factors must be considered, however, to determine if robotics is the right choice for cutting a particular part.

What Does a Robotic Laser Cutting System Consist of?

A typical robotic laser cutting system consists of a servo-controlled six-axis mechanical arm that has a laser "cutting head" mounted to the robot's faceplate. The cutting head provides focusing optics for the laser light and often an integral height control mechanism. An assist gas delivery package is utilized to deliver a cutting gas such as oxygen or nitrogen to the cutting head, which enhances the cutting process. Most systems utilize a Nd:Yag type laser generator, which can deliver the laser light to the robot cutting head through a fiber optic cable. Some systems have been developed to utilize a CO^2 type laser generator, with delivery of the laser light to the robotic cutting head through a mechanical flexible tube arrangement with mirrors at each joint.

Why Use a Robot for Laser Cutting?

Robots offer key features that make them attractive for laser applications.

- **Flexibility to perform 3D cutting.** With a minimum of six axes of freedom, a robot can reach in and around tooling to cut required features into a part. Cut features range from simple round holes to complex contour cutting.

- **Ease of integration and automation.** With the relatively small footprint of the robot, small work cells can be designed to have one or several robot(s) close to a part. Adding one or more robots within the same cell can increase future production. Often two robots are able to share one laser, providing significant cost/benefit factors.

- **Reusable asset.** Because the robot utilizes a flexible programming method, it can be redeployed to cut new parts within the same work cell or moved to a new work cell.

- **Competitive cost/performance.** When compared to conventional hard automation processes such as drilling, punching, or sawing, robotic laser cutting is very competitive. Although the initial capital expenditure is typically higher for a robotic laser cutting system, the improved quality and elimination of multiple workstations yields process and quality return on investment values of two years or less on many systems.

- **Process flexibility.** With advanced controls and some creative application engineering, a robotic laser system can be set up to perform multiple applications, such as cutting and welding with the same laser and "cutting" head.

What Makes a Good Application for Robotic Laser Cutting?

Not all applications for laser cutting should be applied to robots, but many are applicable. Some key factors to consider are as follows:

- **3D parts and closed sections.** Some parts have very complex shapes (i.e., draw stampings) that require the cutting head to achieve many different attitudes to cut required features. Also, parts can have closed sections such as hydroformed tube sections used in automotive frames where holes can't be punched because of access limitations to the back of the material. A robotic laser cutting system is not competitive in high-volume flat sheet cutting that can be performed by a dedicated 2D laser-cutting machine. Nevertheless, if both 2D and 3D cutting is required, the robotic laser cutting system may be applicable.

- **Medium to high volume.** There must be enough part volume to justify the cost of robotic laser cutting equipment and fixtures to hold the part(s). Because a robot is flexible, it can often cut a family of parts in the cell, justifying the cost of the system. Typically, part volumes should be 50,000 to 100,000 or higher, depending on the amount of features to be cut.

- **Aluminum and steel parts.** Most applications for aluminum and steel parts are between 0.5 to 5 mm in thickness for robotic laser cutting. The limitation for cutting thickness is based on laser performance and power. Aluminum requires considerably more power than steel to cut the same thickness. As the power of the laser goes up, there is a significant increase in the cost of the laser generator. Another factor to consider is material coatings such as zinc on steel, which can reduce cutting speed, and anodizing of aluminum, which can improve cutting speed.

- **Fixtures.** Part location must be repeatable and allow access to all areas to be cut. Some parts are very difficult to locate because of a lack of defined features or part holes. Also, holding the part must not interfere with the cutting process, which is often an engineering challenge. These things must be considered when defining the application.

What Capabilities Must Be Provided by the Robotic Laser Cutting System Vendor?

- **Highly repeatable laser cutting robots.** To achieve high-quality cutting, the robot must be capable of locating cut features in the parts to ± 0.25 mm or better.

- **Ease of programming.** Capability needs to be integrated into the robotic laser cutting system to streamline the process of creating and maintaining robotic programs. The system should accept parametric entry of data for the size of standard shapes to be produced such as round hole, slots, and rectangles. Once production cutting is being performed, capability to adjust the feature location and size without interrupting production greatly improves system uptime. Also, the system vendor should be able to create robot programs off-line to reduce programming time and validate the application before physical system build.

- **Process expertise.** The robotic laser cutting system vendor should be able to address all issues regarding the application of the laser cutting process including fixture design, integration into an automated cell with a CDRH conforming light-tight enclosure, and mechanisms to reposition parts.

- **Process control.** Controlling a robotic laser cutting system requires more than just turning the laser on and off. The vendor must be able to control cutting process parameters such as laser power, assist gas pressure, and cutting head height settings. Automated recovery from errors should be integrated into the control scheme to maximize system uptime and prevent costly downtime.

What Should the End User Do to Be Successful?

- **Hands-on approach.** An end user should be involved from the beginning and have intimate knowledge of the overall system, especially the details of the parts to be produced. While the vendor is responsible to develop a system that meets expectations, it is the end user that will work with the system on a daily basis.

- **Process knowledge.** To be successful, the end user needs to become a process expert. One person or a team of people should be identified as the process experts. The team does not have to be process experts before the first system, but should grow into the expert role with training and hands-on experience. The process expert will then be able to maintain and fine-tune the system for optimal performance.

- **Product training.** Training on the robotic system should be planned into the project from the beginning and should be done as early in the project as practical.

Conclusion

Robotic laser cutting is a very useful tool that can be utilized by industry today if applied with proper consideration.

61

■ **Figure 4.14**   *FANUC's new laser-cutting robot.*

How Small Shops Can Stay Competitive with Robotic Solutions

By Kapyoung Choi, Program Manager, FANUC Robotics America, Inc.

As the U.S. economy makes a slow recovery and foreign competition, particularly in China, continues to gain momentum, small shops are realizing that robots are more than just a piece of manufacturing equipment. Robots are key business tools that provide a means for companies to fight back, win orders, and remain profitable. In fact, since the first industrial machine-tending robot was introduced in the United States in the 1960s, the industry has grown to more than 126,000 machine-tending robots in operation (Robotic Industries Association [RIA], 2002).

Many companies now realize that to stay competitive, they must have the manufacturing flexibility to respond quickly to market demands. As companies strive to enhance their time to market, the role of robots becomes particularly critical for smaller shops, where ergonomic issues and absenteeism impact the bottom line.

By implementing robotic solutions, small shops are able to increase their revenue per production employee by 50 percent, and reassign workers to less hazardous and repetitive tasks. Also, robotic automation helps companies streamline operations and realize a quick payback.

The following provides pointers on how small shops can get started on the road to automation.

Understanding Your Needs

An effective automation supplier is one that understands its customers' process needs. By performing a needs analysis to evaluate current manufacturing processes and business priorities, the automation supplier can make recommendations that will provide efficient and cost-effective results.

Because the day-to-day activities of running job shops often prevents production managers from evaluating manufacturing alternatives, it's best to find a local integrator or robot OEM to kick start the needs analysis process with an automation audit. The Robotics Industries Association (RIA) is a good resource to locate suppliers/integrators (http://www.robotics.org). The RIA Web site offers a list of robot suppliers, integrators, tooling suppliers, and other robotic peripheral equipment suppliers.

Advantages of Robotic Automation

Industrial robots are more affordable than just 10 years ago. It may surprise some manufacturers to know that a six-axis robot with a payload of 5 kg, packaged with a six-axis manipulator, controller, and software, is available for less than $30,000. With competitive leasing programs offered by some vendors, small shops can find alternative ways to finance robotic systems at low monthly payments.

In addition, robots are extremely easy to install and operate. Many prepackaged cells are available that allow customers to integrate the robot, tooling, part delivery unit, cell guarding, and control interface to the peripheral devices/machinery in a matter of hours.

For all types of applications, including machine tending, robots are much more sophisticated than earlier models. There are several choices of industrial robots with capabilities to meet a wide range of payload, reach, speed, and flexibility requirements. Typically, robots can handle payloads that range from 3 kg to as many as 600 kg, and offer reach capabilities of 700 mm to more than 3,000 mm. Servo-controlled industrial robots are used to tend CNC lathes, mills, machining centers, drills, grinders, EDMs, and more. They are accurate enough to load three jaw chucks, live tooling, fixtured tombstones or pallets, and collets.

Robots that have at least four axes of motion will help manufacturers optimize cell layout, floor space usage, clip management, and work piece flow. Robots can be mounted on the floor, upside down, on a machine tool, or on a floor track. More recently, a six-axis overhead, rail-mounted robot (Toploader) has gained popularity for tending multiple machines from the top. These robots operate as articulated gantries and allow for more efficient use of floor space and capital.

Speed is another advantage of robotic solutions. The load and unload cycle time for robots is just 5 seconds or slightly more. The time is based on how long it takes for a robot to move into the machine, exchange a part with the machine's work-holding device, and for the machine door to open and close. If the robot supplier designs the system correctly, the robot should wait for the machine, not vice versa.

Picking a Robot that Fits the Application

There are several items to consider when sizing a robot to a machine tending application. The robot OEM or integrator will assign a team that will help evaluate the application during the needs analysis.

■ **Survey parts and group them into families based on size, weight, production requirements, machining time, and machining operation.** First-time robot users should not expect to automate all parts at once. It's best to start with one part family, develop a good understanding of the automation, and gradually automate additional part families. Starting at a slower pace will allow companies to thoroughly evaluate the benefits of robots and streamline the implementation of future automation projects.

■ **Determine the best route for parts to move into and out of the robotic system.** Many part delivery systems are costly; however, the robot supplier will be able to recommend the most efficient and affordable delivery system. How long does the cell need to run unattended? What is the profile of the part or the part family? How much is in the budget? What is the expected return on investment? Answers to these questions will help narrow the choice of part delivery methods.

    – The use of gravity conveyors with part escapement is an effective means of presenting parts to an automation cell and typically less expensive than powered conveyors. Another choice might be part pallets that move manually or automatically into and out of the robot work area.

    – The use of vision with a simple indexing belt conveyor or table is another possibility. Vision provides higher flexibility and reduces the cost of part location. More than one part type can run on the same cell without the need to invest in dedicated fixtures. Vision is more forgiving for slight part-to-part variations compared to a fixtured part delivery system designed to handle nominal part tolerances. When a six-axis robot is used with vision, the robot will position itself at the optimum pickup location based on how the part is positioned for pickup versus dedicated automation, which is designed to pick up parts at the same location each time.

- Vision provides part location and orientation information to guide the robot. Part delivery flexibility will increase if vision is integrated into the robot's end-of-arm-tooling. This allows parts to be delivered to the cell in a structured bin or on a simple multilayered tray system. In addition to eliminating the high cost of part fixturing, vision provides the capability to automate small part runs.

- **End-of-arm-tooling (EOAT) design is defined by part size, weight, gripping location, gripping surface quality, throughput, and work holding device interference zones.** Many grippers used for machine tending applications are off-the-shelf pneumatic parallel motion types with two or three jaws, depending on the part's shape at the gripping location. Others have customized fingers and pneumatic valve systems that control the EOAT. Based on the requirements of the application, some gripper modules are assembled with part-present sensors, vision, and/or part orientation mechanisms.

- **Application software has also been simplified to the point where users with little or no experience can program a robot.** Today's programs are provided in plain English and use drop-down menus to select commands or functions. The teach pendant can be used for programming, to jog the robot, and to monitor and control the robot cell.

  - A single robot can tend one or two machines with the robot set as the master and the CNC machine as the slave. This allows the robot teach pendant and the controller to be used as the operator station, offering simplified panel functions such as graphic status displays, online help and diagnostics, production reporting, and the ability to surf the Web.

  - Local and remote monitoring networks can be created with Ethernet connections on robot controllers. Other controller features include the ability to multitask different activities, perform PLC ladder tasks, detect collision without the use of external sensors, define interference zones between the robot and peripheral devices, and control auxiliary axes.

- **Guarding and safety is an important aspect that should be reviewed and understood by everyone involved in the design, implementa-**

tion, and production of a robotic system. The American National Standard for Industrial Robots and Robot Systems—Safety Requirements (revision of ANSI/RIA R15.06-1986) was approved August 19, 1992, and is available from Robotic Industries Association.

- **Today's robots provide maximum flexibility and are often capable of duplicating human dexterity.** With six axes of coordinated motion and a programmable machine controller, robots can be used to automate an existing or new machine. In either situation, the machine tool must be updated to accept robot automation with features such as an automatic door, automatic work-holding device, physical input/output points to control the external device, and logic in the PMC to interface with a robot.

Industrial robots provide a number of direct and indirect economic benefits. Usually one robot can perform the work of three to five people, reducing the cost of labor in the machine tending area. As the price of robots continues to drop, manufacturers can realize other direct and indirect economic benefits, including:

Direct

- Reduce scrap.
- Lower product liability costs, such as shipping defective parts.
- Increase machine capacity.
- Reduce workers' compensation liability.

Indirect

- Perform postprocessing tasks such as inspection and gauging with one robot.
- Eliminate operator error.
- Reduce machine cycles.
- Increase the possibility of future business.
- Achieve predictable machining processes.
- Improve competitive position.

With the ability to perform a variety of tasks and achieve a mean-time-between-failure of more than 60,000 hours, robots are also more advanced than traditional automation process systems such as linear gantries.

Faced with global competition, manufacturers are reinventing their factories and building manufacturing systems using robotic technology that help produce high-quality products at a reasonable cost. If a company's objective is to achieve one or more of the following competitive advantages, then robotic automation is the right business tool for its success:

- Increase incremental productivity.
- Flexible and predictable production.
- Improve part handling.
- Increase labor savings.
- Eliminate monotonous tasks.
- Provide safer working environment (hazardous environment, back injury, and carpal tunnel syndrome).
- Improve machine utilization (30 percent or more).
- Improve quality (reduce or eliminate the risk of defective parts).
- Reduce work in process and production downtime inventory.

■ **Figure 4.15** *FANUC's LR Mate 200 six-axis robot.*

Automotive Paint Shops: Quality from Technology and Reducing Contamination

Ernest M. Otani, Staff Engineer, FANUC Robotics America

Automotive Paint Shops

Within each automotive assembly plant is a class 8 clean room, where identity and glamour are delivered to sheet metal. Paint and styling are the key identifiers of the consumer vehicle. Together they imply status, glamour, and quality at an emotional level impossible to capture by any sales brochure. Any salesperson or auto show representative can tell you a beautiful and brilliant coat of paint commands attention and helps to make a sale.

The paint shop is responsible for providing this quality paint job to every vehicle while simultaneously managing the cost of the specialized paint materials and an array of environmental regulations for managing air quality. To be successful, the manufacturers must focus on every detail of the process. For just as a quality finish is a major advantage, paint defects are obvious to any observer and taint not just the vehicle carrying the defect, but the reputation of a whole product line.

A successful painting operation is vigilant against contamination and has a very consistent and effective means for delivering paint to the car body. Contamination can come from personnel, equipment, or even the painting process itself. In a typical paint shop, personnel are required to wear basic clean room attire: hairnets, shoe covers, eye protection, and paint suits made from a lint-free fabric. Despite these safeguards, people are probably still the main source of contaminants in the paint environment. By judicious use of quality automation—robots and other painting machines—operators can stay out of the paint booths during production, with only occasional intrusions for service or to clean equipment.

Even painting equipment can be contamination trouble spots if proper care is not taken. A painting robot should be easy to clean—with no features that would tend to trap paint or dirt. Materials exposed to the paint environment should not deteriorate in the presence of harsh paint solvents.

The paint itself may prove to be a contaminant if it builds up on the surface of painting equipment to the point where it might drip onto a job. Many plants use robot coverings or "pajamas" as both a means to manage contamination and as an aid to shorten maintenance time. It is typically much faster and economical to swap a disposable cover than it is to adequately clean equipment with solvent sprayers and rags.

A consistent process begins with paint. Typically, paint is delivered to the plant in 550 gallon containers, with each batch of paint essentially identical to all others of the same color. Nevertheless, a careful paint shop logs in each batch by manufacture date and container number, and takes steps to assure that viscosity and agitation are maintained at manufacturers' specifications. The "mix kitchen," where paint enters the delivery or paint circulation system, is often temperature controlled. All these steps are meant to eliminate any source of variation.

The spray booth itself can affect paint quality by the painting environment it presents. Extreme temperatures and changes in humidity affect the rate at which solvents evaporate from the paint both while airborne and on the painted job. Humidity also affects the effectiveness of the electrostatic charging of the paint—particularly on those waterborne systems that do not charge the paint directly, but use high-potential cathodes that emit ions into the air. Although not all automotive plants control temperature and humidity as closely as other factors, those with less control are faced with the task of fine-tuning their process as these environmental factors shift from day to day and from season to season.

Atomizing Paint

Of course, the most important factors are those related to the specific processing of the job—atomization parameters and the painting path executed by the robot carrying the atomizer. In principle, the job is quite simple; push a uniform cloud of droplets toward the car body in a predictable manner in the allotted cycle time. Need more paint? Increase the paint flow rate. In practice, the quality measures that result in the emotional showroom experience described earlier make the task anything but simple.

Today, atomizers fall into roughly two styles: gun and bell. The spray gun has an atomizing nozzle and fan of air to disperse the droplets in a predictable pattern. The bell has a spinning bell-shaped cup that creates droplets by centrifugal force and an air nozzle that shapes the droplets into a circular pattern directed at the vehicle. The following discussion focuses on bell atomizers because their dramatically higher efficiency makes them the new industry standard.

For bell atomizers, the key parameters are paint flow rate, bell cup rotation speed, and the velocity and shape of the shaping air. For a given flow rate, higher cup rotation or higher shape air velocity lead to smaller particles. And up to a point, smaller particles tend to lead to better quality numbers. But the specifics of how the two factors interact with paint density and viscosity to produce a uniform, brilliant paint job are different for different styles of applicators. And the physics of how the film of paint on the bell cup becomes a uniform cloud of droplets is still an active research area, both in industry and academia.

A final aspect of this problem is the inclusion of electrostatic effects. For optimal painting efficiency, the paint in this cloud of droplets is charged with high-potential (60 kV and higher) ions. The droplets are then drawn to the grounded vehicle body by electrostatic force. This attraction leads to high transfer efficiency—it is common for 90% or more of the atomized paint to actually arrive on the target.

Paint can be charged directly by placing the paint directly in contact with a high-voltage source. Charging the bell cup, or a fluid passage, for example, would achieve this. Paint can also be charged indirectly. Applicators using this strategy emit streams of ions into the surrounding air and have them impinge on the droplets once airborne. This is usually used for conductive paint materials like waterborne paint. The approach has some disadvantages, though, as the cathode "antennae" used to emit the ions can be cumbersome and might make processing more difficult for some jobs. Their many slender rods also present a large surface area, which, if it collects paint, can be an additional source of contamination.

An atomization strategy, electrostatics, and the equipment to produce the desired behavior with consistency and control—all these parameters come together in the spray applicator design. The FANUC Versabell applicator, for example, makes use of a bell rotation speed of 25,000 to 40,000 rpm, in combination with a high flow shaping air nozzle, resulting in a straightforward color matching capability. The geometry of the shaping air nozzle produces a relatively uniform pattern of air that is turbulent enough to further atomize droplets in the cloud. It employs a direct charge approach capable of producing potentials as high as 100 kV.

Challenges in Painting

The objectives of the automotive paint shop are the same as any manufacturing endeavor: quality and efficiency. But competition makes the standards for these objectives very demanding. All glamour and emotion aside, automotive paint quality is measured quantitatively with several specialized metrics. Color is compared to a standard paint sample using an electronic colorimeter. Measurements are taken from several viewing angles because the metal or mica flakes within the paint subtly shade the color as it is viewed from different angles. Film build is often controlled to tolerances in the range of 0.01 mil. And there are several appearance measures for the surface of the finish, including such things as "orange peel" and "distinctness of image" or DOI, which is essentially a measure of the glossy, mirror-like quality of the finish. The metrics offer a clear definition of the quality objective, but meeting that objective with a new or troublesome paint can require extensive experimentation. Once a desired result is achieved, maintaining it requires a good process control plan.

The other key objective is efficiency. This means reduce wasted paint, maximize uptime, and keep maintenance and service overhead to the bare minimum. Here, the right choice in automation can make a substantial difference. A modern painting robot, such as FANUC Robotics' P-200E, is designed specifically for the painting process. Color changing manifolds are located out on the robot arm near the spray equipment to reduce time to change colors and reduce the paint wasted as the paint line is purged and cleaned for the new color. Even performance parameters, such as reach and manipulator motion performance, are tailored to the specific requirements of painting. This kind of automation, which is dedicated to the application, often comprehends practical matters like cable and hose management much more effectively than a general-purpose robot converted to serve the painting application.

The Future

Challenging paint materials are becoming increasingly common. Two-component cyanoacrylate paints have very desirable antichip properties, but must be mixed "on-the-fly," just upstream of the applicator. Waterborne paints have very low levels of environmentally unfriendly volatile solvents, but because the paint is conductive, it presents some additional challenges in isolating the paint to be sprayed from the rest of the paint circulation system. FANUC's Servobell applicator uses a docking station to "fill up" the applicator with a measured charge of paint that is then separated from the circulation system and sprayed onto the vehicle. Direct charge of the waterborne paint eliminates the need for the external array of cathodes carried by indirect charge systems.

In the coming years, a number of trends will drive the next generation of painting equipment and processes, including:

- Emergence of "special handling" paints as mainstream.
- Further advances in productivity for paint atomizers and robot arms.
- Bell style applicators with traditional atomizing spray guns as the equipment of choice in the automotive industry.
- Improved understanding of atomization physics, leading to even higher quality measures with less setup and experimentation.
- Environmental concerns, leading to a shift toward water-based paints and more demands on air handling systems in the paint booth.

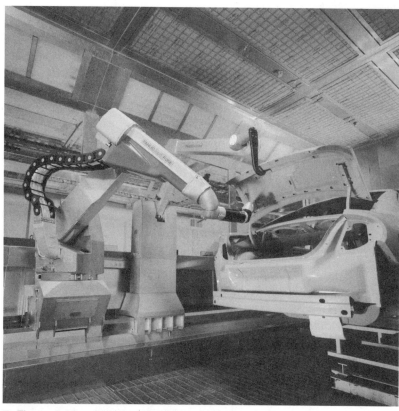

■ **Figure 4.16** *FANUC's P-200E painting robot.*

Kuntz Electroplating Develops a New Automated Wheel Polishing Process

By Martin Step, Kuntz Logic Systems, Inc.

Kuntz Electroplating Inc., of Kitchener, Ontario, is one of the world's largest independent OEM chrome plating suppliers, producing millions of plated automotive wheels, bumpers, and other components annually. Kuntz is a family-owned business that has been in operation for over 50 years, employs over 1,000 people at peak production, and is an approved supplier to the Big Three, Harley-Davidson, and many others.

In recent years, a trend in new wheel styles often includes deeply contoured spokes. Traditional, fixed automation polishing and buffing machines cannot reach the deeply recessed areas between the spokes of the wheel. To accomplish the polishing operation, hundreds of finishing workers are required to work in a 24/6 manual operation.

The manual finishers polish the wheels with sandpaper cartridge rolls mounted on die grinders. The subsequent chrome plating operation requires that the finish on the wheels be of exceptionally high quality. Any areas that are not blended or smoothed properly will be very noticeable once the plating operation is complete. Finding, training, and retaining hundreds of manual finishing personnel became a daunting task. Manual polishing is tedious and monotonous, and the average new worker lasted less than a month. Also, these polishing operators accounted for more than half of the entire company's repetitive strain injuries.

Because of these difficulties, it became apparent to Kuntz Management that some type of automation was required. After conducting a survey of available technology, it was determined that Kuntz would have to develop its own automated process for wheel finishing. Kuntz has extensive experience in designing and implementing fixed automation for automotive wheel finishing, but it was determined that fixed automation would still not be effective for the newer wheel styles with deeply contoured spokes. Therefore, a completely new approach was needed for this application.

Kuntz contacted FANUC Robotics at an early stage of the project, in search of an off-the-shelf solution. Dr. Laxmi Musunur, manager of the Material Removal group at FANUC Robotics, had completed many robotic finishing and deburring systems, but had never attempted to polish aluminum wheels with such stringent quality requirements.

After some discussion, FANUC Robotics was able to help Kuntz select the appropriate robot and process equipment, and several things became evident. A variable speed (up to 6,000 rpm), high torque motor, capable of automatic media change and a compliant tool would be needed. PushCorp, Inc., in Dallas, Texas, was chosen to provide a critical component in the end-of-arm tooling. Its active force-compliant tools would provide the compliance and accurate forces necessary to duplicate the human touch of the manual finishing operation.

Initial test results were very promising, and Kuntz decided to install twenty M-710i robots. This was a very ambitious project, with ten cells of two robots each. The cell design featured two inverted robots mounted on an elaborate steel superstructure. The superstructure was situated above a large turntable capable of fixturing ten wheels.

Following start-up and debug, the first of these cells began producing wheels in mid-1999, with the last cell being completed about 1 year later. At this point, a very difficult issue arose. Robot paths for the first three wheel styles were programmed by manual lead-by-touch methods, and each wheel path program was taking 8 to 10 weeks to create.

The lengthy programming time was due to the extreme complexity of the wheels, one of which had fifteen spokes. There was also a considerable learning curve to overcome in trying to understand the fine points of the human operator's techniques and reproduce them adequately. The goal was for the robots to achieve the same smooth, blended finish characteristic of wheels that had been manually polished.

To address this problem, it was decided that some type of off-line robot path programming software was needed. To acquire this capability,

Kuntz purchased a Krypton RODYM-6D Robot Measurement System and contracted the services of Al Knasinski, of Logic Systems Development Inc. This combination of equipment and personnel allowed the development of an off-line programming (OLP) solution based on using a 3-D CAD model of the wheel. These techniques promised to give better control over all of the parameters, such as force head pressure, spindle speed, travel speed, and blending paths, to produce smoother, more consistent paths in less time. The system could also be used to calibrate the robots and easily transfer programs from one cell to another.

Work began in earnest utilizing the robot measurement and programming system to develop robot programs for the new model-year wheels. Unfortunately, this effort brought to light a significant shortcoming in the cell design that was not apparent with the original 1998 model-year wheels.

The 1998 wheel styles were shaped such that the sandpaper cartridge orientation relative to the wheels did not deviate much more than about 10° from vertical. However, in 2000, Kuntz began working on developing paths for the 2001 model-year wheels. It soon became apparent that there was a fundamental problem with the cell layout. The 2001 wheels had a significantly different spoke shape that required that the sandpaper cartridge be able to reach positions at a much shallower angle relative to the wheel face.

Even after a costly conversion of the wheel fixtures to reposition the wheels, a number of other issues became apparent. The new robot path programs required much faster motions, which caused flexing in the robot support superstructure. This led to surface defects in the polishing when one robot's motion would cause the other to vibrate. Also, the larger motions required by the new wheel styles often forced the robots into singularities.

After making a concerted, but ultimately futile effort to make the original cell design work for the new wheels, Kuntz decided to reconfigure the cells into a totally new design, which was referred to internally as the "Second Generation." In the new cell layout, the robots were mounted conventionally, upright on a pedestal. In front of each robot are two programmable wheel-mounting platters, which are the robot's seventh and eighth auxiliary axes. FANUC Robotics came to Kuntz's assistance by supplying one of its M-6*i* robots.

As the robot polishes, the wheel is rotated about its own axis, allowing coordinated motion with the robot. Because the wheel is continually repositioned, the robot works within the dexterous part of the work envelope. While the robot polishes a wheel on one auxiliary rotary axis, the operator can unload and load the other unit, permitting continuous operation. The two auxiliary axis units are separated by safety fencing; each has pressure safety mats and a complete set of operator controls. There is an elaborate system of safety interlocks on the auxiliary rotary axis units and on the robot to ensure that the operator can safely load one unit while the robot works on the other. This is not a trivial consideration because the unit the operator is loading is still "live" and under control of the robot controller throughout the duration of the operator's exposure!

Throughout the effort to get the original cells performing properly and development on the Second-Generation concept, other significant developments were underway. The first was a business decision by Kuntz Electroplating Inc. to form a new division called Kuntz Logic Systems Inc. (KLS). The new division was intended to develop off-line programming solutions, robot metrology, and integrated systems. The business unit was formed by purchasing the assets of Logic Systems Development Inc. from its majority owner, Krypton Electronic Engineering of Belgium. The asset purchase was completed in late 2000, and KLS officially began operations on January 1, 2001.

The basic concept of OLP is not new, but as many people in the industry are aware, the Achilles' heel of OLP is that it has been very difficult, if not impossible, to achieve a simulation-created robot path that requires no real-world touch-up. Because the polishing paths necessary for automotive wheels are among the most difficult robot programs, the task was extremely challenging.

To polish automotive wheels, there are numerous complex and interrelated factors to coordinate. Of primary importance is that the working axis of the PushCorp-compliant tool must always be normal to the surface at the point of contact. Also, the motor travel speeds, compliant-tool forces, and motor RPM are in a constantly varying balancing act dependent on the local surface geometry. So, if any one factor must change, the others must be automatically adjusted to suit. Acceleration and deceleration must be carefully controlled at all times. Initial contact and liftoff points must be feathered to avoid polishing discontinuities. Also, wear patterns on the sandpaper cartridges must be controlled for a long, even polishing life. The sandpaper cartridge lead and toe-in angles must be constantly adjusted for smooth blending of polishing marks. In short, it is nearly impossible to develop such complex paths by manual lead-to-teach methods. With some styles of polishing media, there is a considerable change in tool diameter as the media wears down, which also has to be compensated for by the program.

Another complicating factor is the inconsistency of the lug nut hole locations relative to the wheel spokes. The wheels are positioned by these holes; but, according to the manufacturer's specifications, the placement of these holes can vary rotationally by ± 1°. On an 18-inch (457 mm) diameter rim, this leads to an error of almost 1/4 inch (8 mm) near the outer edge of the wheel. This can cause major problems with robotic polishing. KLS subsequently developed a proprietary method for automatically detecting and compensating for this error that does not require the expense of a vision system.

Last, the OLP system has to be capable of quickly creating different versions of any one wheel polishing path. This is necessary to be able to cope with the varying surface quality of the incoming raw cast aluminum wheels. When the surface conditions deteriorate, the robot operator needs to have the option of adding additional polishing paths, or even different grit sizes of sandpaper to the base routine from the robot's control panel. This means that the programs must be capable of adapting on-the-fly to changing requirements without having to create or load new programs.

The process of starting a completely new business division and subsequently developing an OLP package is a very complex and time-consuming task for a chrome plating supplier. However, at the time, Kuntz management believed that this was the best solution. Several companies supply OLP packages, but each suffer from various limitations that severely limited their usefulness to Kuntz Electroplating.

The task of developing OLP solutions for polishing wheels starts by loading a 3-D CAD model of the wheel into Robocad™. Next, the wheels are subdivided into various portions based on their geometry, and detailed subroutines are developed to handle the particular requirements of each kind of geometry. Then, for each wheel, a master program assembles the subroutines in the correct order, depending on the number of passes selected by the operator for each different grit size. At this point, a certain level of intelligence is added to the system to semi-automate the process of assembling the master program. This application is referred to as automatic path generation (APG).

Another very significant piece of software developed for this project is a new interface between the simulation software and FANUC robot controller software (RCS), which KLS calls the "LS_RCS Interface." All of the new KLS software runs as a Robocad open system environment (ROSE) function in Robocad™, which is a robot simulation software package produced by Tecnomatix Technologies, Inc. Like all robot simulation software packages, Robocad contains a robot motion planner; however, a robot path generated with simulation software may not run the same way when it is downloaded to a robot. This is because the real robot controller software is different in many respects. Our solution was to develop an interface that allows the programmer to select the actual FANUC RCS (made available separately by FANUC Robotics) to do all the motion planning. In fact, this new interface allows the full use of every function contained within the RCS, which immediately opens up a much wider range of programming tools to a simulation engineer.

There are two other essential elements used to ensure a touch-up free download to the robot. The first is another KLS application—an OLP interface that ensures 100 percent round-load integrity. This means that there is absolutely no loss or degradation of data in a robot path transferred from simulation software to the robot controller or vice versa. The second essential element is that the target robot polishing cells are fully calibrated using a RODYM-6D™ Robot Measurement System produced by Krypton Electronic Engineering of Belgium. The calibration procedure compensates for any dimensional irregularities between the simulated robot cell and the real one. One additional piece of KLS software called VitualTeachPendant, which is very helpful but not necessarily essential, is a teach pendant emulator, which lets the OLP operator view, modify, and run the robot path in simulation exactly as on a real teach pendant.

KLS has now reached the point where a polishing path for a new wheel can be developed by a skilled programmer in approximately 1 to 2 weeks using all of the tools developed to date, including testing and optimization. This compares very favorably to the 8 to 10 weeks that it took for lead-to-teach paths in the original polishing cells. In addition, the automatically generated paths are far superior in terms of smoothness of motion and overall improvement of wheel quality.

Obviously, this approach to OLP has broad implications aside from polishing wheels. For example, both FANUC Robotics and a major automaker are using the new RCS interface as part of a software package supplied by KLS for off-line programming of paint robots.

Additional work is presently at the research stage, including a major R&D project at the University of Waterloo, which should further reduce the interval. This project was set up with the assistance of a number of companies. PushCorp of Dallas, Texas provided its latest

Active Force Head and Spindle. FANUC Robotics Canada, Ltd. supplied an M-710*i* robot and controller. Kuntz Electroplating Inc. provided all of the remaining hardware, and is donating a complete Krypton RODYM-6D Robot Measurement System. Just as the new RCS interface has broad implications beyond aluminum wheels, it is expected that the work being done in the university R&D project will move OLP to the next level. One of the first areas of research is to develop the means to automate the task of optimizing the robot paths. This would eliminate a great deal of the trial and error in testing various robot polishing paths.

By late 2001, Kuntz built a prototype Second Generation Pilot Cell consisting of two robots and four auxiliary axis units. The new system produces extremely consistent work and has exhibited none of the problems that plagued the first generation design. The quality of the output is nearly indistinguishable from that of the best human polisher. After an extended trial, the remaining robots will be reinstalled in the new configuration.

The ability to quickly program complex automotive wheels provides Kuntz Electroplating with a unique manufacturing advantage. These cells can obviously be used to run continuously on large production runs, but they also offer a very advantageous short-run capability. Many automakers are now looking at more frequent changes of trim components, such as wheels, as a means of offering special editions and for keeping their overall product offerings fresh. Also, by diversifying its customer base beyond the Big Three, Kuntz expects to considerably increase the number of short runs.

With large product runs of 10,000 wheels or more per week, the manual operators are able to hone their proficiency, resulting in greater throughput and improved efficiency. However, with short runs, some as low as 500 or less per week, this level of proficiency never develops. If short-run wheels can be polished on a robot, then the quality level and throughput will not depend on achieving and maintaining operator proficiency.

To this end, Kuntz has been using its Robot Lab development cell to test and develop the means to incorporate the other types of polishing media in addition to the sandpaper cartridge. These media may include small flap wheels, brushes, and sandpaper spinners of various styles. The goal is to be able to eliminate all hand polishing for any wheel style that requires these processes. The first wheel to be programmed for robotic polishing using all of these types of media as required was a wheel for a major automaker's SUV, which is a very large, complex surface area to be polished. The polishing quality achieved by the robotic process was found to equal or exceed that of the best human polishers, but at only half to one-third of their typical time.

As this book is going to press, Kuntz is finalizing plans for reinstalling up to twelve of its robots in this new Second-Generation configuration, to be ready for full production in late 2002. This new line will handle all of the polishing for a major new wheel program slated to launch in January 2003. This line will have an added twist, in that it will also incorporate fully automated loading and unloading of the auxiliary axes units, so that only one operator will be needed to run all the robot polishing cells.

■ **Figure 4.17** *FANUC's M-710i robot for automated wheel polishing process.*

Robotic Dross Removal— Skimming More Profits for the Hot Dip Steel Galvanizing Industry

What do you get when you take a zinc metal bath, heat it up to 878° Fahrenheit, throw in the backbreaking job of manual dross skimming, and run production 24 hours per day, seven days per week? You get all the justifications you need for a great robotic application.

That is exactly what one of the world's leading hot dip galvanizers of automotive coiled sheet steel found after it began the production use of a FANUC Robotics Model M-170*i* robot to perform the laborious task of dross skimming in January 1999. According to Mark Habros, Health and Safety Coordinator at DNN Galvanizing of Windsor, Ontario, backaches, muscle pains, and lost time associated with manual dross skimming has been eliminated, thanks to the addition of an industrial robot to its production team.

Prior to robotic dross skimming, DNN Galvanizing had annually rotated employees through the task of skimming the dross (metal oxides) that form on the 6 foot by 13 foot surface of the zinc galvanizing bath through which they dip more than 147,000 lbs. of sheet steel per hour. Habros said, "It is a hot, physically demanding job that has to be done almost continuously to maintain the quality of the zinc bath." Dross skimming requires an operator to manipulate a skimmer (a large "spoon" with holes in it attached to a 10-foot-long handle) across the entire surface of the zinc bath in multiple passes. The skimmer stops frequently to lift up to 33 pounds of dross that accumulates on the tool, and drops the dross into a small holding pot for off-site recycling.

■ **Figure 4.18** *Dross skimming on a zinc metal bath.*

■ **Figure 4.19** *Manual dross skimming.*

Ambient air temperatures over 100° Fahrenheit are not uncommon near the molten metal bath. The possibility of burns from splashed molten metal are a constant danger of manual dross skimming, requiring workers to wear additional protective clothing, making the job even hotter and more uncomfortable. The balcony over the zinc bath creates a low overhead clearance, and the close proximity of the overhead rail for the removal of dross pots makes this a particularly confining work area for a manual operator.

Beyond the demanding physical requirements of dross skimming, there is a tedious and repetitive nature about the job that can cause a manual operator's pace to slow and the quality of his or her work to become inconsistent by the end of the shift. This decrease in quality at the dross skimming operation can have a direct impact on the paintability of the sheet steel at the automotive end user's final assembly plant. In 1998, DNN Galvanizing Team looked for an automation solution to meet the many requirements of the dross skimming challenge.

DNN's search for a total turnkey robotic dross skimming system led them to FANUC Robotics. DNN contacted FANUC Robotics to review the requirements and provide a proposal. After the careful evaluation of all bids, FANUC Robotics Canada Ltd. of Mississauga, Ontario was contracted to provide an automation solution consisting of:

- M-710*i* six-axis industrial robot, all-electric, servo motor driven, with model R-J2 controller.
- HandlingTool™ application software package.
- Stainless steel robot end-of-arm-tool (EOAT).
- Two-position manual slide for the robot mechanical unit.
- Removable molten metal splash cover for the two-position slide and robot base.
- Wire mesh safety enclosure with interlocked gate.
- Project management, engineering, and manufacturing.
- Classroom training on the robot and controller.
- Complete on-site robot start-up and programming.
- On-site supervision of initial production run.
- On-site system training.
- System documentation.

The M-710*i* robot is programmed to make multiple passes across the surface of the zinc bath similar to the way the manual operator did the job. After every pass, the robot lifts the skimmer out of the zinc bath and makes a motion just above the surface of the metal bath to move the skimmer closer to the dross pot. During this motion, the liquid zinc passes through the holes in the skimmer, falling back into the zinc bath and leaving the dross in the skimmer. The robot makes a short move from the edge of the metal bath to a position over a dross pot located beside the robot. The robot skims the entire surface of the zinc bath within approximately forty skimming motions in a period of about 15 minutes. In a typical 12-hour shift, the robot fills an average of six dross pots.

Dross pot replacement requires a temporary shutdown of the robot because the operator must enter the robot's work envelope to remove the full dross pot and replace it with an empty one. The operator selects the shutdown sequence, which sends the robot to a safe stopping position after completion of the current skimming motion and emptying of the skimmer. A safety interlock on the gate to the robot cell automatically puts the robot into an Emergency Stop whenever the gate is opened.

■ **Figure 4.20** *FANUC's M-710i robot showing the hot, dirty environment where an operator is replaced.*

The only maintenance required for the robot is the replacement of the portion of the skimmer tool that comes into contact with the molten metal and dross. It is typically replaced once every 6 months due to erosion of the skimmer.

After 9 months of successful robotic dross skimming, Habros said, "At the time of the installation, our employees were skeptical of the robot; they were concerned the robot wouldn't be able to perform as good a job as an operator." But after a short time, the employees became comfortable with the successful robotic skimming operation.

Typical of many robotic operations that are tied to in-line processes, many nonrobotic variables exist that can require manual modification of the robot's program to ensure consistently successful operation. Dross viscosity and the amount of dross on the surface of the zinc bath varies from day-to-day. Manual dross skimmers make up for that variability with changes in how often they go to the dross pot to empty their skimmer. The robot has no way of detecting changes in dross viscosity and layer thickness. Therefore, minor edits are keyed into the robot's teach pendant through user-friendly menus to slightly adjust the robot's program for dross variations. Additionally, the robot cannot see when dross has frozen to its skimmer, requiring the skimmer to be immersed and cleaned in the hot zinc bath until the frozen material melts off. The robot has a regular routine to perform this immersion of the skimmer. An unscheduled "skimmer cleaning" can also be initiated with the simple push of a button by the operator who oversees the robot cell.

Some areas of the metal bath are completely out of the robot's reach, such as the 1 foot by 13 foot area on the opposite side of the sheet metal delivery snout. To assist the robot with these areas, a manual operator does a small amount of dross pushing with a skimmer, moving the inaccessible dross to locations where the robot can reach it. This same operator has the occasional task of directing the nozzle of a long hose to inject nitrogen gas along the edges of the metal bath where dross is prone to freezing prior to being robotically skimmed.

DNN has logged a reduction in lost time due to the elimination of physical problems associated with manual dross removal. The robot has freed the former skimming operators to do other tasks that are less strenuous, physically taxing, and tedious. DNN has not been able to quantitatively measure the improvement in employee morale from the dross skimming robot. However, it is certain that the feedback of the employees who work with and around the robot is the best measure. DNN employees are said to be happy with the newest member of their production team, the Dross Skimming Robot, and they speak positively about the job it performs. In a manner of speaking, the dross removal robot is skimming up more profits for them and their company by performing the most labor-intensive, backbreaking, and monotonous job in their facility.

DNN Galvanizing opened its hot dip galvanizing and galvanneal line (HDGL) in Windsor, Ontario in 1992. NKK Iron and Steel Engineering Ltd. (NKK-SE) supplied the HDGL process equipment. The company is a joint venture and is operated as a limited partnership managed by a general partner, DN Galvanizing Corporation, coating steel for Dofasco and National Steel on a toll-coating basis. The venture is owned 50 percent by Dofasco, 40 percent by NKK, and 10 percent by National Steel.

FANUC Robotics North America, Inc. designs, engineers, and manufactures innovative robots and robotic systems that impact business performance for global customers. FANUC Robotics, headquartered in Rochester Hill, Michigan, is a robotics industry leader. A subsidiary of FANUC Ltd. in Japan, the company has facilities in Chicago; Los Angeles; Charlotte, NC; Cincinnati, Ohio; Toledo, Ohio; Toronto; Montreal; Mexico City; Aguascalientes, Mexico; and Sao Paulo, Brazil. For more information on FANUC Robotics, call 1-800-47-ROBOT or visit www.fanucrobotics.com.

Robots Impact Production at Mennie's Machine Company, Inc.

At Mennie's Machine, the term "innovation" is not a buzzword; it is business as usual. The Illinois-based manufacturer of machined castings and forgings now uses robotic automation to meet its customers' stringent demands for speed and quality. Previously a manual process, human limitations had prevented Mennie's from achieving throughput and consistency requirements for a major contract to manufacture automotive drive shaft components for sport utility vehicles and full-size trucks. In addition, the company had safety and ergonomic issues for employees, due to repetitive activities. Consequently, frequent employee absenteeism made production planning nearly impossible.

Operating three shifts per day, the company used a manual operation for one year before Mennie's determined that automation was needed. The manual system, which typically yielded 80 pieces per hour, did not achieve production goals of 118 pieces per hour.

"We realized that in order to position ourselves as a world-class manufacturer, we needed to implement leading-edge technology," said Mennie's vice president Bill Mennie. "On the project, speed and quality were the highest priorities because when it comes to automotive manufacturing, there is no margin for error."

According to Mennie, production complexity was another key factor in the decision to go with robotics. In addition to the many different processes and types of machinery involved in the production of finished parts, the company was challenged to balance cycle times between each process. With the manual operation, it became difficult to overcome bottlenecks in the system, preventing Mennie's from achieving its desired production goals. Mennie's selected FANUC Robotics North America, Inc. for a turnkey solution because of FANUC Robotics' reputation for having reliable products and systems expertise. In addition, Mennie's appreciated having a supplier within close proximity to its manufacturing facility. Although FANUC Robotics is headquartered near Detroit, its full-service facility, located just outside of Chicago, took complete responsibility for the design, build, installation, and service of the system.

Mennie's has a system design philosophy of: "Factory within a factory." This meant that it needed a series of robotic work cells linked through a common part transfer conveyor. "FANUC Robotics understood what we were trying to accomplish and encouraged an atmosphere of teamwork," said Mennie.

The fully automated system incorporates nine material handling robot models. Mennie's provided the pallets, part fixtures, safety fencing, conveyor equipment, PLC hardware, and other peripherals.

Each robot is equipped with HandlingTool and Collision Guard software, custom grippers, and a control interface to the equipment and conveyor. The HandlingTool software package provides built-in functions, menu-driven prompts, and point-and-shoot position teaching, making it easy for operators to create and run programs. Collision Guard allows a robot to sense a potential collision along any axis and stops robot operation in time to minimize damage to the end-of-arm-tooling, the part, or the robot.

The entire system consists of ten consecutive processes that are linked via conveyor. Raw parts are first manually loaded into a "face and center" machine before being conveyed downstream through nine automated machining processes. At each "stop" along the conveyor, parts are automatically loaded and unloaded by robots. First, an M-6i, capable of lifting up to 13.2 pounds, picks raw parts and places them into a CNC lathe for machining. The robot then places finished parts onto a part pallet that is located on the conveyor. Next, each full part pallet makes its way to an LR Mate 100i tabletop model robot, which loads and unloads parts into and out of a small broach machine. In the next process, another M-6i alternately loads and unloads parts onto one of two fixtures of a large broach machine for further processing. A third M-6i moves parts into and out of a boring operation.

In the final processes, an S-430*i*F, which can lift up to 286 pounds, sequentially loads and unloads parts into and out of four machines that perform various groove and ream operations. Following this operation, parts are loaded and unloaded by a fourth M-6*i* robot to and from a drilling operation, then conveyed to another LR Mate 100*i* for a spline-roll machining process. Next, an M-16*i* robot, capable of lifting up to 35 pounds, handles parts two at a time into a heat-treat operation. Finally a fifth M-6*i* is used to load and unload parts for grinding.

At Mennie's, production quality is assessed by tracking scrap dollars. During final inspection, each part attribute is examined for dimensional tolerance and cosmetic appearance. Rejects are tagged and then further evaluated to determine how to eliminate inaccuracy. Robotic automation has contributed to a reduction in scrap rates. In addition, as a result of the robotic system, Mennie's has experienced significant competitive advantages including:

- Accurate and consistent part loading.
- A reduction in part defects.
- Increased productivity—typically 120 pieces per hour.
- Improved control of the entire manufacturing system.
- Less cost per piece.
- Flexibility to meet future production demands.

The close working relationship established between Mennie's and FANUC Robotics was a major factor in the success of the project. "Our expectations were met and responsibilities were understood without experiencing any unexpected surprises," said Mennie. "I credit open and honest customer/supplier communication for that." Another contributing factor to the successful system was the commitment that Mennie's had to using automation to achieve its goals. "We put our trust in FANUC Robotics' engineering team to supply a turnkey solution that would help us achieve our vision of 'factory within a factory,' and they didn't let us down," said Mennie.

Production is humming at Mennie's. The three-shift operation is running at 95 percent uptime, a 15 percent increase over the previous manual operation. In fact, Mennie's estimates that the robotic system has helped the company reduce total production costs by approximately 25 percent. Robotic automation has also helped Mennie's increase in production flexibility. Not only is the system expandable to accommodate additional machines, but with slight modifications to fixtures, the system could easily accommodate new part designs. Another flexible feature of the system is that Mennie's can switch to manual production during routine preventive maintenance. Since the original robotic cells were installed in 1997, Mennie's has expanded the system twice, including the purchase of an additional robot for a drilling operation and an additional machine tool to increase throughput in a turning operation.

Mennie's Machine Company is a privately held Illinois corporation specializing in low- to high-volume subcontract manufacturing for the heavy construction and off-road equipment and automotive OEM markets. Founded in 1970, by Hubert and Cheryl Mennie, the company is QS 9000 and ISO 9002 certified and has achieved approximately $25 million in annual sales. For additional information, contact Mennie's Machine Company at 815-339-2226 or visit its Web site at www.mennies.com.

■ **Figure 4.21** *FANUC's LR Mate 100i robot for loading and unloading broach machine.*

Industry—Roller Bearings

Year installed—1998

System Overview

An IRB 1400 robot is used for loading and unloading of the sensor-bearing parts from a drill machine. A mechanical (three-jaw/finger) gripper is utilized for handling one part at a time. The gripper has sufficient stroke to be able to grip any one of four part outer diameter (ODs). New unprocessed parts arrive positioned, stopped, and located on the infeed conveyor for robotic unloading. Processed parts are robotically placed onto the exit conveyor.

Parts

There are four sizes of sensor bearings made of steel.

Customer Benefits

Customer benefits include:

- Manpower reduction.
- Reduced manual intervention.
- Improved part quality.
- Increased machine uptime.
- Reduced cycle time.
- Flexibility of production process.
- Continuous operation.

- Improved equipment utilization due to reduced machine idle time.
- Reduced changeover time.

Unique Elements

The unique element of this application is that a robot is used in place of a standard pick and place unit of fixed automation to give greater flexibility in part handling.

Major items to be furnished by the robot manufacturer were:

- The IRB 1400 Robot.
- The Robot Gripper.
- The pneumatic valve package.
- The interface to the drilling machine.

Customer-provided equipment was the drilling machine, the infeed conveyor, the exit conveyor, and the safety equipment.

Project/Steps Provided

Project/steps to implementation include: concept formulation, proposal engineering, project management, mechanical engineering, electrical engineering, system assembly and test, system programming and debug, electrical design and engineering, manufacturing and build, system installation supervision, and documentation.

93

■ **Figure 4.22** *ABB's IRB-1400 robot for loading and unloading sensor bearing from a drill machine.*

■ **Figure 4.23** *ABB's IRB-1400 robot that handles four different sizes of bearing diameters for loading and unloading.*

Industry—Aluminum Die Castings and Transmission Housings

Project name—Shotblast unload cell

Year installed—1998

Parts—Aluminum transmission cases

System Overview

Transmission case castings are exiting a shotblast chamber via a wire mesh belt. Orientation is variable due to forces generated during blasting. The orientation to the transmission location case is registered by the vision system; the robot adjusts and picks the part from the moving conveyor. Excess shot is dumped from part cavities, and the casting is deposited on a gravity conveyor, which exits the cell.

Customer Benefits

Customer benefits include:

- Reduced manual intervention for handling consistency, improved personnel utilization, and the elimination of part damage.
- Ergonomic enhancements: reduced back and hand injuries (carpel tunnel).
- Improved part quality.
- Reduced cycle time.

- Integration of other plant equipment.
- Continuous operation.

Unique Elements

The unique element of this system is that it used an Optimaster Vision system in a new and unique manner.

Major Equipment Provided

Major items furnished by the robot manufacturer include:

- IRB 6400F Foundry robot.
- Pneumatic actuated gripper.
- Camera/lighting mast.
- Pneumatic valve package.
- Optimaster Vision processor.
- Robot riser.
- Outfeed conveyor.
- System perimeter guarding.
- Planned cell entry gate box.
- Interface for shotblast system.

Customer-provided equipment includes a shotblast system.

Project Steps to Implementation

Steps to project implementation include:

- Concept.
- Proposal engineering.
- Project management.
- Mechanical engineering.
- Electrical engineering.
- System assembly and test.
- System programming and debug.
- Manufacturing/build fabrication.
- System installation supervision.
- Documentation.

■ **Figure 4.24** *ABB's IRB 6400 F foundry robot.*

■ **Figure 4.25** *ABB's IRB 6400 F robot for loading and unloading aluminum transmission housings from shot blast chamber.*

Industry—Automotive Assembly Seam Sealing of Cars

Date installed—July 1994

Project Description

The one-cell system consisted of two ABB IRB 3200 servo track-mounted robots to do selected interior seam spraying on the floor pan, and one IRB 3200 overhead-mounted robot to do roof ditch seams. Four vision sensors mounted underneath the vehicle and three overhead are used to locate the vehicle. Nordson Pro-Flo dispensing equipment was incorporated into the new system supplied by existing bulk supply.

Customer Benefits

Customer benefits include:

- Reduced labor.
- Improved quality (lower warranty claims).
- Consistent, repeatable application of sealant materials.

Unique Elements

Unique elements include:

- Jobs per hour—80.

- Total cycle time—43.5 seconds (longest vehicle).
- Bead size—3 mm +2/−0 high, 25 mm +3/−0 wide.
- Material spec: 252.
- Total length of sprayed seams—359.5 inches (approximate).
- Average dispense speed—10 inches/second (per robot).
- Maximum dispense speed—20 inches/second (per robot).

Major Equipment Provided

Major equipment supplied includes:

- Three IRB 3200 robots, controllers, and dispense end of arm tooling (EOAT).
- Overhead mounting structure for the roof ditch robots.
- Two IRBT 3002 serve tracks.
- Nordson Pro-Flo dispense system.
- Material temperature conditioning systems.
- Material filter modules.
- Seven perception vision sensor devices with mounting structure.

Project/Steps Provided

Steps to project implementation include:

- Concepts.
- System design.
- Engineering.
- Manufacturing.
- Project management.
- Application processing and robot programming.
- Installation supervision.
- Documentation and training.

Industry—Personal Care

Project name—Turnkey palletizing system

Date installed—July 1998

Parts—Cases of tablets

Process Description

100

The following is a description of the process:

- Cases are conveyed into the system on single conveyor.
- Bar codes on cases are scanned.
- Products are diverted onto one of eight infeed conveyors, based on SKU.
- Misreads are diverted out of the system.
- If the infeed is full, product recirculates, which allows for buffer and accumulation.
- Robot picks up a row of cases with vacuum and places cases on one of four pallets.
- Robot can form a column stack or complex patterns.
- Full pallets are automatically loaded onto the track car and conveyed to stretch wrapper.
- Stretch wrapper wraps the load and discharges stable pallets from the system.

Unique Elements

A turnkey complete palletizing system in a combination of two IRB 640 high speed palletizing robots, state-of-the-art product scanning, high-speed case diverters, a versatile end effector that can pick up any product with no changeover, pallet handling conveyors, track car, stretch wrapper, and state-of-the-art integrated control system with man–machine interface underscore the crucial advantages the system has to offer.

This system offers flexibility of action, agility of production, fast and easy product changeover, and very low operating and maintenance costs, with superior uptime.

It has championed the concept of speed with flexibility. It is a complete, comprehensive, and tightly integrated turnkey system that is simple and inexpensive to operate and that surpasses the flexibility and reliability associated with "traditional" palletizing systems.

Equipment Provided

Equipment provided by the robot manufacturer includes:

- Two IRB 640 FlexPalletizer® robots.
- Universal vacuum end effectors.
- Product supply conveying system.
- Product identification and sorting system.
- Product recirculation and accumulation conveying system.
- Pallet/slip sheet conveying system, including track car.
- Pallet stretch wrapping system.
- Complete comprehensive electrical and control system with central man–machine interface.
- Perimeter guarding system.

Customer equipment interfaced to robots includes:

- Eight case packing systems on different floors in the plant.
- Forklift removal of unitized loads at output of system.

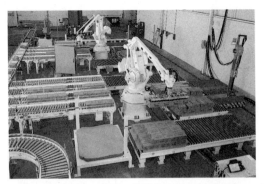

■ **Figure 4.26** *ABB's IRB 640 flex palletizer robots.*

■ **Figure 4.27** *ABB's IRB 640 turnkey palletizer system.*

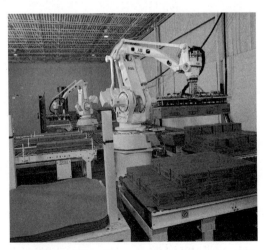

■ **Figure 4.28** *BB's IRB 640 palletizer robots that can handle eight different products simultaneously.*

Industry—Mail Processing Industry

Year installed—National deployment of up to 450 systems by 2000.

Parts—Individual trays or tubs filled with mail weighing 15–70 lbs.

Customer Benefits

Customer benefits include:

- Palletizes up to 24 different ZIP coded trays at ten to fifteen trays per minute.
- Standard configuration consists of two twelve sort-position modules (other configurations available).
- Randomly processes different trays and tubs.
- Zero missort.
- Eliminates need to manually lift 15 lb. trays or 70 lb. tubs.
- Palletizes into mail containers and onto flat pallets interchangeably.
- High equipment uptime during container/pallet exchange.
- Ease of changeover to different palletizing schemes.
- Production database provides user configurable reports.
- Computer interface to upper level diagnostic maintenance system.

Project Description

Trays and tubs of mail are transported past a bar code scanner and go to one of two pickup positions. The bar code is communicated to the system controller, which directs where one of the twin gantry robots palletizes the tray or tub of mail.

Highlight Elements

Unique elements of the system include:

- Complete turnkey system.
- Graphical user interface, controls, and host computer interface.
- Long-term training and service support.

Major Equipment Supplied

Major equipment supplied includes:

- Gantry Robot with gripper.
- Conveyor system.
- Standardized modular infeed to meet site-specific requirements.
- Operator interface and system controls.
- Safety system.

■ **Figure 4.29** *ABB's Gantry robot and conveyor system.*

■ **Figure 4.30** *ABB's Gantry robot and conveyor system.*

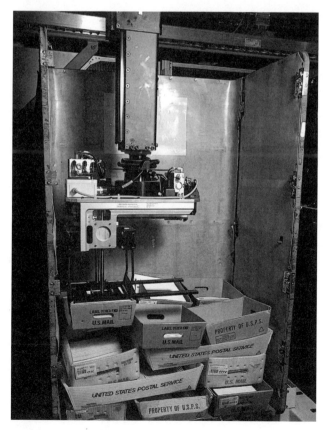

■ **Figure 4.31** *ABB's Gantry robot and conveyor system.*

Industry—Bottling and Packaging

Date installed—October 1993

Parts—Five bottle sizes in common tray

Customer Benefits

Customer benefits include:

- Reduced labor costs.
- Random product handling.

Project Description

Conveys pallets of bottle trays to robot. Robot removes top cover and then individual trays. Trays and bottles are picked up and bottles are dumped down chutes into filling machine equipment. Empty cardboard tray is then placed on belt conveyor.

Highlight Elements

Unique elements of the system include:

- End effector searches for tray sides to compensate for variable tray locations.
- System handles five product sizes without changeover.

Major Equipment Provided

Major equipment provided includes:

- IRB 6000 Robots (4).
- Pallet stacker.
- Full pallet infeed conveyors.
- Bottle dump chutes.
- PLC code.
- Operator station.
- Tray handling end effectors.
- Lexan safety fencing.
- Common empty pallet and empty tray conveyors.

■ **Figure 4.32**   *ABB's IRB 6000 robots for pallet stacking.*

18

Industry—Tobacco Industry

Date installed—August 1993

Parts—10 to 60 lb. cases

Customer Benefits

Customer benefits included:

- Palletized random case sizes.
- Can palletize up to thirty-six different products at one time.
- Eliminates handling of 60 lb. case by hand.
- Verification to ensure products on pallets are not mixed.
- Automatic motion of empty and full pallets (shuttle cars).

Project Description

Project includes providing six robots used for palletizing forty-one different case sizes. Each robot palletizes on one of six 48" × 48" pallets. Cases are delivered on a conveyor to pickup positions where the robot mechanical gripper picks up the case by gripping the top and bottom of the package.

Highlights of the System

Highlights of the system include:

- Complete system provided to the end user and installed in conjunction with equipment provided by three other vendors.
- Dynamic allocation of product and six pallet positions per robot.

Major Equipment Provided

Major equipment provided includes:

- One case conveyor (approx 450 ft.) with seven swing arm diverters.
- Remote input/output and controls for Allen Bradley pyramid integrator. Controls consisted of "real-time" allocation, case tracking, and system diagnostics.

■ **Figure 4.33** *ABB's palletizing robots for handling up to thirty-six different products at one time.*

Industry—Automotive Assembly

Project name—Seam sealing of truck cabs

Date installed—June 1994

Parts—Pickup trucks and sports utility vehicles

Customer Benefits

Customer benefits include:

- Reduced labor.
- Improved quality (lower warranty claims).
- Automotive backup.
- Consistent, repeatable application of sealant material.

Project Description

This is a two-cell, dual-line system consisting of six ABB IRB 3200 robots. Each cell contains two servo track-mounted robots to do interior seam spraying, one overhead inverted mount robot to do the roof ditch seams, high speed indexing conveyor tables, and a vehicle location station. Existing Nordson dispensing equipment was incorporated into the new system.

Highlight Features

Highlights of the system include:

- Jobs per hour—52.
- Total cycle time—69.3 seconds (longest vehicle).
- Bead size—3 mm + 2/−0 high, 25 mm + 3/−0 wide.
- Material spec—252.
- Total length of sprayed seams—15,240 mm.
- Average dispense speed—320 mm/second per robot.
- Max dispense speed—600 mm/second per robot.

Major Equipment Provided

Major equipment provided includes:

- Six IRB 3200 robots, controllers and dispense end-of-arm tooling (EOAT).
- Overhead mounting structures for the roof ditch robots.
- Four IRBT 3002 servo tracks.
- Nordson Pro-Flo Dispense System.
- Material temperature conditioning systems.
- Crash detection devices for the EOAT.
- Rapid transfer conveyors with lift and locate stations.
- Allen Bradley PLC with panel view master operator station.
- Safety fencing and photo cell intrusion detection devices.

Industry—Automotive Subsupplier

Project name—Adhesive dispensing and headlamp assembly

Date installed—December 1991

Parts—Headlamp

Customer Benefits

Customer benefits include:

- Increased production.
- Consistent quality.
- Quick changeover.
- Reduced labor.
- Reduced waste of expensive adhesives.

Project Description

The system utilizes an IRB 2000 GL robot to apply urethane adhesive to a headlamp reflector; then a specially designed pick and place unit assembles the lens to the reflector. There are four fixtures mounted on a dial index table. An operator manually loads two lenses and two reflectors. After the process is complete, an operator then unloads two finished assemblies.

Highlight Features

Features of the system include:

- Increased productivity and improved quality of parts.
- System doubled manual production.

Major Equipment Provided

Major equipment provided includes:

- IRB 2000 GL Robot.
- Cantilever pick and place unit.
- Dial index table.
- Adhesive dispensing equipment.

- **Figure 4.34**   *ABB's IRB 2000 GL robot for adhesive dispensing and headlamp assembly.*

21

Industry—Automotive

Project name—Flexible assembly of cylinder heads

Date installed—June 1992

Parts—Cylinder head (System II)

Customer Benefit

The benefit of this system is that it reduces inspection requirements.

Project Description

The system includes complete assembly of cylinder heads. Major operations are: seal/seat assembly, valve cycling, air test, parts dump, head rotation for valve insertion, spark plug torque down, valve key lock assembly, camshaft seal assembly, valve lock assembly inspection, rocker arm torque, and camshaft insertion.

Highlight Elements

The highlight of the system is the robot key-up operation.

Major Equipment Provided

Major equipment provided includes:

- Two IRB 1000 robots.
- Two IRB 2000 robots.
- Two IRB 3000 robots.

- 11 automated stations.
- Five manual stations.
- Rectangular loop conveyor (14' × 100').
- 25-foot repair bay.

■ **Figure 4.35** *ABB's robot system for flexible assembly of cylindar heads, 2 IRB 1000 robots, 2 IRB 2000 robots, and 2 IRB 3000 robots.*

Industry—Custom Plastic Molders

Project name—Router trimming and hole drilling of plastic panels

Date installed—August 1992

Parts—Cover for truck box tailgates

Customer Benefits

Customer benefits include:

- Greater reliability and output than from previous automation.
- More flexibility for multiple programs.

Project Description

The robot performs final detail trimming and hole drilling on two parts at one station while the operator unloads two fixtures at the other station.

Highlight Elements

Highlights of the system included:

- Replaced five-axis Cartesian NC machine.
- Robot system has much improved throughput, operator load/unload access, and safety.
- Operators can edit existing programs or create new programs, which could not be done with the NC machine.

Major Equipment Provided

Major equipment provided includes:

- IRB 3000.
- Electric high speed spindle and control.
- Pivoting fixture tables.
- Safety barrier with clear Lexan.

■ **Figure 4.36** *ABB's IRB 3000 robot for router trimming and hole drilling of plastic panels.*

Industry—Off-Road Equipment

Project name—Flexible gear manufacturing

Date installed—February 1990

Parts—Planetary gears (146 different parts)

Process Description

Batch manufacturing of gears. Robots are used to handle gears between machine tools, gauges, and washing stations. A cell control computer directs the traffic within each cell. Product between cells is guided and regulated by a higher level controller.

Customer Benefits

Customer benefits include:

- Eliminates "work in process" inventory.
- Reduces frequency of manual intervention during gear manufacturing process.
- Integrates manufacturing of gears into plant-wide control system.
- Reduces manpower.

Unique Elements

Unique elements include:

- Cell and area level software.
- Interface/communications to all machine controllers.

- Installation supervision.

Equipment Provided

Equipment provided includes:

- Sixteen inverted IRB 3000 robots.
- Communication between all devices.
- One area gantry.
- Cell level software.
- Seventeen custom grippers.
- Area level software.
- Nine cell control computers.

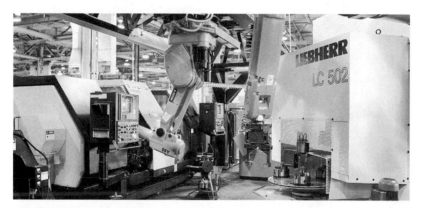

■ **Figure 4.37**   *ABB's IRB 3000 robots for batch manufacturing of gears.*

■ **Figure 4.38**   *ABB's IRB 3000 robots for batch manufacturing of gears.*

■ **Figure 4.39** *ABB's IRB 3000 robots for batch manufacturing of gears.*

Industry—Off-Highway Farm Tractors

Project name—Urethane dispensing system for tractor windows

Year installed—1998

System Overview

An ABB IRB 6400 industrial robot dispenses cold-applied urethane sealant onto tractor windows positioned in front of the robot by an ABB servo track. ABB project content included single-source responsibility for the entire system including both manual and automatic urethane dispensing equipment, as well as installation and integration into an existing assembly line.

Customer Benefits

Customer benefits include:

- Replace highly variable manual dispensing technique—eliminates leaks and/or rework.
- Eliminate extremely messy manual process.
- Reduce manual intervention.
- Ergonomic enhancements—glass is difficult to handle during manual application.
- Improve part quality through consistent urethane application.
- Random processing of one or the other sets of windshield glass.

- Consistent with goal to improve assembly line speed by 30 percent—reduced cycle time.
- Continuous operation.

■ **Figure 4.40** *ABB's IRB 6400 robot dispenses cold urethane sealant for tractor windows.*

■ **Figure 4.41** *ABB's IRB 6400 robot showing guarding and ABB servo track.*

Industry—Aerospace (Helicopters)

Project name—Helicopter gear deburring cell

Year installed—1998

Parts—Helicopter gears (machined steel)

System Overview

An ABB IRB 2400 robot is installed in an enclosure, along with a rotary table to hold fixtured gears for deburring and a tool rack for holding up to ten deburring tools. The operator opens the door at the end of the enclosure, loads a gear into the chuck, exits the cell, and closes the door, and then initiates operation of the robot. The robot starts its operation by picking up the selected tool from the robot program. The robot then positions the tool in front of a fiber optic sensor to either measure the diameter of the tool or to detect breakage, and then initiates deburring of the gear. When the robot finishes deburring, the robot positions clear of the part, stops its cycle, and a beacon illuminates to indicate that the gear is complete for operation exchange of the parts. The operator unloads the finished gear and loads a new gear to be processed.

Customer Benefits

Customer benefits include:

- Labor reduction (savings).
- Eliminates hazardous process during handling of heavy gears (i.e., burns, splinters, metallic dust inhalation).

- Reduces manual intervention (improper loading—resulting in part damage).
- Ergonomic enhancements—eliminates long reaching and back injuries when handling heavy gears, carpal tunnel.
- Improved part uniformity and quality via repeatable robot process and motion.
- Eliminates "work in progress" inventory through consistent, predictable operation.
- Reduced cycle time by 90 percent.
- Provides continuous operation through breaks, lunch, and across shifts.
- Replaces a manual deburring task that is difficult to do uniformly, and is tedious, dirty, and gritty.
- Improved safety measures using planned cell entry procedures.

Unique Elements

Unique elements of the system include:

- Fully integrated seventh axis (part of robot's S4 control system) rotary table minimizes gear tooth deburring programming.
- Fiber optic sensor for deburring wheel diameter measurement and program offset allows use of wheels as they wear.
- Fiber optic sensor for broken tool detection.
- Automatic tool changer system.
- Tool rack with up to ten deburring tools and brushes.
- All equipment mounted on a base, which can easily be moved together to a different site and placed into production with minimum delay and without excessive readjustment.
- Programmable valve for controlling compliance pressure per tool on-the-fly as the robot program runs.

■ **Figure 4.42** *ABB's IRB 2400 robot helicopter deburring cell.*

■ **Figure 4.43** *ABB's IRB 2400 robot shown deburring helicopter gear.*

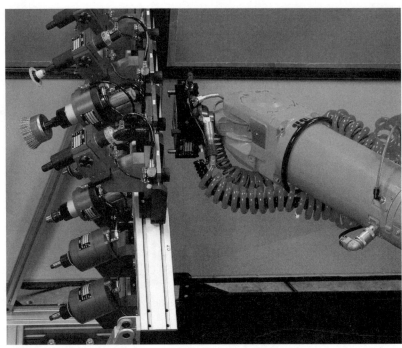

■ **Figure 4.44** *ABB's IRB 2400 robot with tooling rack to hold up to ten gear deburring tools.*

Industry—Motorcycle

Project name—Finishing heat shields (three robot cells)

Year installed—1998

Parts—Heat shields, eleven types (sheet metal)

System Overview

The part processes consist of belt grinding and flap wheel finishing in cells 1 and 2, and belt grinding, flap wheel finishing, and buffing in cell 3. Each robot and corresponding process machinery sits on a common base plate. Parts will be presented to the robot in each cell by one ABB 180-degree index table that has two part fixtures at each diameter end. Each part fixture can hold at least eight to ten parts of the same part number. All part fixtures are supplied in pairs and with quick attach features on the indexer top. This lets the robot pick parts inside the cell and the operator load/unload parts at the other fixture safely, outside the robot work zone. The robot gets a single part from the fixture, performs the process, and then returns this part again to the same fixture. When all parts in the fixture are done by the robot, the indexer turns 180 degrees and presents the next fixture that has already been prepared and loaded with new parts by the operator. The finished parts are presented back to the operator. The operator replenishes the fixture with new unprocessed parts.

Customer Benefits

Customer benefits include:

- Reduces manpower—labor savings.

- Eliminates hazardous process—grinding dust inhalation, back injuries, carpal tunnel.
- Reduces manual intervention (improperly applied grinding and polishing pressure, i.e., part damage and improper finish).
- Predictable media wear.
- Ergonomic enhancements—eliminates carpel tunnel, constant vibration, and noise.
- Improved part uniformity and quality via repeatable robot process and motion.
- Reduced cycle time.
- Throughput increase.
- Minor part changeover.
- Continuous operation.
- Flexibility of production process.

Unique Elements

Unique elements of the system include:

- Automatic tool changer on robot to process different parts automatically.
- Consistency of part finishing.
- Three cells provided.

- **Figure 4.45**   *ABB's IRB 4400 robot with ABB 180 degree index table for heat shield finishing.*

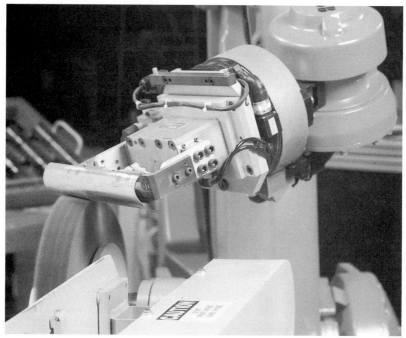

■ **Figure 4.46**   *ABB's IRB 4400 robot showing belt grinding and flap wheel finishing.*

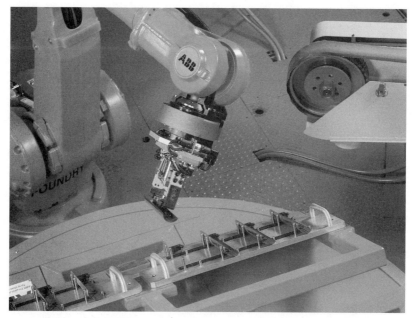

■ **Figure 4.47**   *ABB's IRB 4400 robot showing belt grinding, flap wheel finishing, and buffing on heat shields.*

Industry—Motorcycles

Project name—Finishing sheet getal gas tank

Year installed—1998

Parts—Gas tanks, two sizes (sheet metal steel)

System Overview

Parts enter and exit the cell in the fixtures located at the operation station. The robot picks up the gas tank and takes it to two single-arm belt sanding machines to grind the welded seam on the tank. Following the operations, the robot then takes the gas tank to flap sanding wheel for final finishing. And, when selected by the operator, the robot will brush the tank filler neck with a wire brush wheel on a fourth machine, also located in this cell.

Customer Benefits

Customer benefits include:

- Manpower reduction.
- Eliminates hazardous process—grinding dust inhalation, back injuries, and carpal tunnel.
- Reduces manual intervention.
- Ergonomic enhancements.
- Improves part quality.
- Reduces cycle time.
- Throughput increase.
- Flexibility of production process.

- No part changeover.
- Continuous operation.

■ **Figure 4.48** *ABB's IRB 4400 robot for finishing sheet metal gas tank for motorcycles in enclosed cell.*

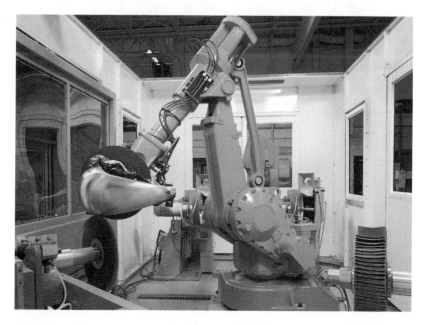

■ **Figure 4.49** *ABB's IRB 4400 robot shown with flap sanding wheel for finishing welded seam on gas tank.*

■ **Figure 4.50**   *ABB's IRB 4400 robot shown with two single arm belt sanders for rough finishing of welded seam on motorcycle gas tanks.*

Industry—Motorcycle

Date installed—December 1997

Parts—Green shafts for motorcycle transmissions: four sizes of parts (two short and two long)

System Overview

A robot machine tends to green shafts on spline roller and gundrill machines.

Customer Benefits

Customer benefits include:

- Increases production by 80 percent.
- Consistent quality.
- Reduces in-process inventory by 50 percent because all part numbers can be processed randomly.
- Reduces labor by 60 percent.
- No changeover for parts required.

Process Description

The system consists of two cells: the spline roller cell and the gundrill cell. In the spline cell, the shafts are spline rolled and either transferred out of the cell through a washer or transferred to the gundrill cell. At the gundrill cell, the shafts are spline rolled, gundrilled, inspected, and transferred out through the washer.

Unique Elements

The system automatically handles four part numbers without operator intervention and uses customer-standard part baskets as transport fixtures.

Major Equipment Provided

Major equipment provided by ABB includes:

- Two IRB 440 robots.
- Basket infeed conveyor.
- Transfer conveyor system.
- Queue stand.
- Undrill cell inspection slide.
- Custom engineered spline cell robot gripper.
- Spline cell inspection slide.
- Custom engineered gundrill cell robot gripper.
- Light inspection station.
- Peripheral guarding.

Customer equipment tended or interfaced to robots includes:

- Traub Model TMC-65D6 horizontal lathe.
- DeHoff (4) spindle gundrill machine.

■ **Figure 4.51** *ABB's IRB 4400 robots (2) tending of green shafts on spline roller and gun drill machines.*

Industry—Power Generation

Project name—Machine tending of connecting rods

Date installed—Summer 1998

Parts—Connecting rod for engine driving generator

Process Description

Connecting rods enter the cell in dedicated dunnage. Robot 1 unloads the two-piece rod and loads a dedicated transfer pallet. The conveyor system delivers the rod to robot 2. This robot handles just the cap portion and delivers it to a lathe, a drilling machine, and a post-process gauge. This gauge records all the features machined and gives tooling offsets to the lathe, as appropriate. The cap is returned to the same transfer pallet for delivery to robot 3. This robot handles the rod portion. When robot 3 delivers the part to the lathe, the matching dimension for the cap is downloaded so the rod cam can be machined for a match fit. Robot 3 also services a milling machine and a post-process gauge. This gauge records the features machined in this cell and gives tooling offsets to the lathe. Robot 3 returns the rod to the mating cap on the transfer pallet. The set is delivered back to robot 1. This robot removes the part from the conveyor, loads a pin press and marker combination machine, and then loads the completed part set into dedicated dunnage for shipment from the cell.

Customer Benefits

Customer benefits include:

- Reduced unload/load cycle time.

- Produces rods in a matched set.
- Keeps rods and caps as a set.
- Posts process gauges for size control.
- Reduces manpower.
- Part characteristic data storage.
- Serial number for part traceability.
- Cell control from host computer screens.

Unique Elements

Unique elements of the system include:

- Three unique gripper designs.
- Serial communication to host comput.
- Pin press machine.
- Marker for permanent part identification.
- Special conveyor system with transfer pallet tracking.
- Work piece data collection for part traceability.
- Offset passing for process control.

Major ABB deliverables and services include:

- Giddings and Lewis Lathes (4).
- Detroit Precision Gauges (2).
- Special Machines Moehrle, Inc. (2).
- Part marker—GT Schnide (1).
- Transfer conveyor—H&CS (1).
- Rod conveyor—Caterpillar (1).

■ **Figure 4.52** *ABB's IRB 6400 robots (3) for tending of connecting rods unload station.*

■ **Figure 4.53** *ABB's IRB 6400 robots (3) for tending connecting rods drilling and gauging operation.*

■ **Figure 4.54** *ABB's IRB 6400 robots (3) for tending connecting rods feeding a lathe drilling machine and a post process gauge.*

Industry—Aluminum (Automotive) Wheels

Project name—Wheel deburring system

Year installed—1998

Parts—Two forged automotive wheel configurations

138

System Overview

An ABB IRB 4400 industrial robot picks up a burred wheel on the incoming conveyor. The robot orients the spokes of the wheel with a laser sensor and positions the located spokes to a stationary-mounted ABB high-speed deburring tool to remove burr, which was produced by an upstream lathe process. Following deburring, the robot deposits the wheel on the output conveyor system.

Customer Benefits

Customer benefits include:

- Replaces highly variable manual technique and irregular intervention, which reduces equipment utilization.
- Ergonomic enhancements—manual deburring difficult to do consistently.
- Improved part quality and eliminates rework.
- Batch processing of one or the other wheels.
- Consistent with production requirements.
- Continuous operation.

Unique Elements

Unique elements include complete cell contained on a base for ease of installation and relocation at a later date. Customer had only to connect air and power before going into production.

Major ABB deliverables and services include:

- IRB 4400 industrial robot.
- ABB high-speed air-driven deburring tool.
- Fully enclosed system mounted on a base, shipped, and installed as an assembly system.
- System installation, power up, and commissioning.

■ **Figure 4.55** *ABB's IRB 4400 robot for aluminum auto wheel deburring system.*

■ **Figure 4.56** *ABB's IRB 4400 robot offloading wheel from conveyor.*

■ **Figure 4.57**   *ABB's IRB 4400 robot using ABB high-speed deburring tool.*

■ **Figure 4.58**   *ABB's IRB 4400 robot shown loading wheel on output conveyor.*

Industry—Plastic Injection Molding

Project name—Panel sanding cell

Year installed—1998

Parts—Injection molded plastic office furniture panels: Various panels including doors for cubicle overhead filing cabinets and file drawer fronts. A total of seven parts were programmed.

Process Description

- Eliminates "work in progress" inventory through consistent, predictable operation.
- Reduces cycle time.
- Continuous operation through breaks and across shifts.
- Replaces a manual sanding task that is difficult to do uniformly, tedious, dirty, gritty.
- Improves safety measures using planned cell entry procedure.

Customer Benefits

Customer benefits include:

- Labor reduction (savings).
- Eliminates hazardous process during sanding (e.g., dust inhalation).
- Reduces manual intervention (improperly applied sanding pressure—resulting in part damage and improper finish).

- Ergonomic enhancements—eliminates long reaching and back injuries when sanding panels, carpal tunnel, constant vibration, and noise.
- Improves part uniformity and quality via repeatable robot process and motion

System Overview

An ABB IRB 4400 robot is installed in an enclosure, along with a rotary table for holding two fixtures. While the robot is sanding a part(s) on one fixture, the operator exchanges a part(s) on the other fixture.

Unique Elements

Unique elements of the system include:

- Compliance device mounted on end of robot arm for maintaining constant pressure during sanding.
- Compliance device with programmable pressure can be changed on the fly via the root program.
- Operator menu system allows the operator to select the number of sanding passes, the pressure of each sanding pass, and inclusion/exclusion of selected sanding areas.
- Common fixturing system for two families of parts.
- Automatic sanding tool changer: Stand is supplied with up to four sanding tools, which are exchanged automatically when an operator has selected the number of the parts to be processed.

Major ABB deliverables and services include:

- IRB 4400 robot.
- Pneumatic valve package.
- Pneumatics panel.
- Common base structure for robot and rotary table to maintain alignment over time and ease of installation.
- Compliance device for sanding pressure.
- Perimeter barrier with interlock-controlled gate access.
- Universal fixturing system for two families of parts.
- Sanding tool stand for up to four sanding tools.
- Four sanding tool assemblies.
- Operator panel for rotary table.

■ **Figure 4.59** *ABB's IRB 4400 robot in panel sanding cell of plastic injection molded furniture panels.*

■ **Figure 4.60** *ABB's IRB 4400 robot showing close-up of one station.*

32

Industry—Textile Manufacturing

Project name—Flexible material handling and palletizing

Date installed—December 1993

Part—Spools/Bobbins

Customer Benefits

The benefit of this system is that it relieves operators of highly difficult, physically demanding, and error prone packing of spools.

Process Description

- Packing spools in cartons and finishing cartons by assembling sides and top.
- High variety of spool type and carton configuration as well as frequent changeover of product to be packed characterizes the system.
- Uses the Optimaster vision system, which allows the customer to continue using existing overhead power and free carriers (approximately 400 carriers). This saves the customer from purchasing either new carriers or reworking existing carriers.

Unique Elements

Complex and extensive cell software was developed to meet textile industry needs. During project execution, ABB was able to accommodate project changes without schedule delay. These changes increased the scope of the original plan and enabled the customer to plan for additional future capacity requirements.

Major equipment provided by ABB includes:

- System cell controller.
- Three-servo track.
- Optimaster vision systems.
- Three IRB 6000 robots.
- Carton material handling system.
- Overhead power and free system with car track and conveyors.

145

Industry—Textile

Project name—Flexible material handling and palletizing

Date installed—May 1993–August 1993

Part—20 lb. spools of nylon

Customer Benefits

Customer benefits include:

- Multiple merges (mix) are identified, labeled, and packed.
- Sorting/combining of multiple qualities of products optimizes production.
- Packages are automatically rejected for operator inspection/reintroduction.
- 100 percent verification of type/quantity for all packages.

Process Description

- Project includes three identical packaging complexes, each containing three robots.
- Each complex is designed to deliver the required throughput while providing maximum flexibility for strategic material handling and grouping of the product.

Unique Elements

Unique elements of the system include a complete turnkey, which includes:

- Preproject development.
- VAX level control.
- Cell and system simulation.
- Installation.
- Interfacing with existing equipment at customers' site.
- Temporary setup and test.

Major equipment provided by ABB includes:

- Nine IRB 6000 robots.
- Twelve slip sheet feeders.
- Nine O.D. multi-package grippers.
- Eighteen storage racks.
- Three belt conveyors with transfer car.
- Six package identification stations.
- Three 3100 VAX computers and Allen Bradley PLC5-40s.

Diesel Engines for Heavy Tractors and Trucks

Customer—Caterpillar Tractor

This case was taken from the current files of a Systems Integrator, Cinetic Automation, 23400 Halsted Road, Farmington Hills, Michigan 48335. Contact Jeff Donovan 248-477-0800.

Part—Six-cylinder diesel engine (approx. 700 lbs.)

Task—To set the timing (valve lash)

Machine vision—DVT infrared LEDs

Robots—Three large ABB robots need to handle the weight

Procedure

Assembled engines appear on a J-hook conveyor completely assembled except for valve covers. This condition is called "ready for cold test." One robot rotates crank to top dead center (TDC), and the other robots set and lock lash-on valves with dual spindle nut-runners, one to set position and one to tighten lock-nuts while three gauges check position. LED illumination is critical because very little space is available with an assembled engine.

PLC—Allen Bradley with diagnostic software by National Instrument. All programming and final software by Cinetic Automation, as well as safety light curtains and support system design.

System was shipped late in 2003, but due to minor changes in the engine, will probably be operational in mid-2004.

■ **Figure 4.61** *ABB's valve lash fixture.*

■ **Figure 4.62** *ABB's IRB 6400 robots for setting the valve timing on caterpillar tractor diesel engine.*

References

1. Lindbom TH. Today's Robots at Work in Industry: Matching the Robot and the Job. *Proceedings of Second International Symposium on Industrial Robots.* May 1972:130.

2. Hangawa Y. An Approach to Industrial Robot Application Research. *Proceedings of Second International Symposium on Industrial Robots.* May 1972:10–11.

3. Lindbom TH. Matching the Robot and the Job. *Proceedings of Second International Symposium on Industrial Robots.* May 1972:129.

4. Hangawa Y. An Approach to Industrial Robot Application Research. *Proceedings of Second International Symposium on Industrial Robots.* May 1972:12.

5. Anderson A. Private interview held during a question and answer session of IEEE Seminar on Industrial Robots at Lawrence Institute of Technology, Detroit, Michigan, May 1972.

6. Lindbom TH. Robots: Capabilities and Justification. *Manufacturing Engineering & Management.* July 1972:17–19.

150

Upkeep of Industrial Robots

5

Design and Construction of Industrial Robots

The design of an industrial robot is critical to the satisfactory application to the job to be done, as well as its long life in production, with minimal maintenance. Successful robot manufacturers must have rugged machine members with precision axes and ways. They must include rugged part grippers that will not damage the part, yet will give many hours of production with little wear. The most critical item of all, however, is the software to go with the hardware. The major cause of production line downtime where robots are involved is the software program designed to control the robot. Those who are familiar with the "lockup" of their home computer can empathize with this problem. We seem to be able to train technicians to solve problems with the arms, grippers, and power supplies with little difficulty, but technicians who can solve programming problems quickly are in very short supply. "Nissan, in their $1.43 billion plant in Canton, Mississippi, hired and trained 3,000 people as production technicians to work in the plant's body, paint, trim, and chassis shops. But 5 months after the plant opened, Nissan is still struggling to find industrial maintenance technicians, workers who are critical to keep the lines running. They're responsible for 17 miles of conveyor belts, 853 robots, 70 lasers, and 300 programmable controllers that control the entire assembly process."[1]

Setup

The setup of a robot is just as important as the design, in that it can be an excellent design, but if it is not set up properly, results can be disastrous. The setup should take into account the speed and time of each movement, as well as the end-of-arm tooling adjustments. Part orientation as presented to the robot, as well as part orientation upon leaving the robot, is critical to smooth operation. Well-trained setup personnel are critical to a zero-defect type of production.

Maintenance

Maintenance personnel should be well trained in the mechanical adjustments of the robot, as well as the settings for electrical and electronic parts of the robot. They must know the program software because the entire system is dependent on many things working together. In other words, they must have a broad background in many disciplines so that they can anticipate problems before they happen, keeping the lines running with little or no stoppages. These kinds of people are hard to find. It has been the author's experience that only about one in fifty can have the requisite skill level for proper robot maintenance.

Reference

1. "Nissan seeks techs to keep line moving." Barbara Powll (Associated Press). *Detroit Free Press*. October 10, 2003, Section C, pages 1, 3.

Economic Justification

6

The economic justification for an industrial robot has been explored in three different ways by Lindbom: payback period, return on investment, and present value of future earnings.[1]

Payback Period

The basic equation is:

$$P = \frac{I}{L - M}$$

where P is the payback period in years, I is the total investment for the robot and accessories, L is the annual labor savings, and M is the annual maintenance expense for the robot. Typical inputs are: I = \$25,000, L = \$11,000, M = \$2,000 for one shift use or \$3,000 for two-shift use.

Substituting in the equation,

For one-shift operation:

$$P = \frac{\$25,000}{\$11,000 - 2,000} = \frac{25,000}{9,000} = 2.7 \text{ years payback}$$

For two-shift operation:

$$P = \frac{\$25,000}{\$22,000 - 3,000} = \frac{25,000}{19,000} = 1.3 \text{ years' payback}$$

The preceding basic equation assumes that the robot and the human can operate at nearly the same rate of production. In the case where this differs substantially, the equations may be modified accordingly.

$$P = \frac{I}{L - M + q(L + Z)}$$

where P is the payback period in years, I is the total investment for the robot and accessories, L is the annual labor savings, M is the annual maintenance expense for the robot, q is the production rate coefficient, and Z is the capital value of the equipment served by the robot (typically 15 percent of acquisition cost).

Typical inputs are: $I = \$25,000$, $L = \$22,000$ for two shifts, $M = \$3,000$ for two shifts, $q = 20\% + 0.20$, and $Z = (15 \text{ percent}) \times (\$200,000) = \$30,000$.

Substituting in the equation for two-shift operation in which the robot is either 20 percent faster or 20 percent slower than a man:

When faster:

$$P = \frac{25,000}{22,000 - 3,000 + 0.2\,(22,000 + 30,000)} = 0.85 \text{ years' payback}$$

When slower:

$$P = \frac{25,000}{22,000 - 3,000 - 0.2\,(22,000 + 30,000)} = 2.9 \text{ years payback}$$

Return on Investment[2]

$$R = \frac{\text{Average Annual Cash Flow}}{\text{Net Investment}} = \frac{\text{Annual Labor Savings} - \text{Annual Maintenance Costs} - \text{Annual Depreciation}}{\text{Net Investment}}$$

$$\frac{\$4.75/\text{hr} \times 16 \text{ hr/day} \times 5 \times 50 - \$5,000}{\$25,000} = \frac{14,000}{\$25,000} = 56\%$$

Where an investment of $25,000 is assumed as the net investment for robot and accessories, straight line depreciation for five years is $5,000 per year, maintenance is $0.75 per hour of operation, an average labor rate with fringe benefits is $5.50 per hour, and a two-shift, 16-hour day, 5 days per week, and 50 weeks a year is assumed.

Present Value of Future Earnings[1]

If it is determined that an acceptable rate of return is at least 20 percent and that life is 5 years, then present value can be calculated by using a factor of 3.256, obtained from a Present Value table for investments. Using values from the previous two paragraphs, it is calculated that a robot working two shifts and earning a net of $19,000 per year is worth as much as $61,800 today.

Present value on a one-shift basis:

P.V. = 3.256 × ($11,000 − $2,000) = $29,300

Present value on a two-shift basis:

P.V. = 3.256 × ($22,000 − $3,000) × $61,800

Comparison of These Three Measures

In comparing payback, return on investment, and present value, one may gain added insight by analyzing a typical example of an automation solution to an assembly problem. Following is a summary of the pertinent data:

	A Robot A	B Robot B	C Fixed Automation
Net investment	$24,000	$35,000	$60,000
Average annual operating cash flow	14,000	16,000	48,000
Life of investment	4 yr.	5 yr.	10 yr.
Total operating cash flow during life	56,000	80,000	480,000
Payback	1.7 yr.	2.2 yr.	1.3 yr.
Return on investment	58%	46%	80%
Present value of future earnings	39,500	52,000	218,500

However, to determine a common denominator, we must first analyze each alternative to determine what adjustments must be made so that each option provides the same amount of goods or services, in this case, the same production rate required. In this instance, the Fixed Automation Option provided a much faster production rate, and it was necessary to have three robots (A or B) to match its production. The new comparison chart would then read:

	A Robot A	B Robot B	C Fixed Automation
Net investment	$72,000	$105,000	$60,000
Average annual operation cash flow	42,000	48,000	48,000

Life of investment	4 yr.	5 yr.	10 yr.
Total operating cash flow during life	168,000	240,000	480,000
Payback	1.7 yr.	2.2 yr.	1.3 yr.
Return on investment	58%	46%	80%
Present value of future earnings	118,500	156,000	182,000

In this case, it is obvious that the Fixed Automation alternative is the desirable one, but it is equally apparent that care must be used in selecting the parameter with which to make this judgment. For example, suppose that cash flow was to be our only criteria for selection. Then, because options B and C both produce the same cash flow, either could be chosen. However, the payback, return on investment, and present value all favor option C. In the case of comparing options A and B, A has the best payback and return on Investment, but the present value favors option B.

The MAPI Formula

The Machinery and Allied Products Institute (MAPI) has pioneered in the development of formulas applicable to investment decisions. Dr. George Terborgh, the research director of the institute, published a volume in 1949, discussing the theory of equipment replacement.[3] More recently, the formula has been refined and developed in a form more easily understood by the practical businessperson.[4]

The MAPI formula seeks to take into consideration other factors that tend to have impact on the automation investment decision:

- The formula should make some kind of projection of technological change, that is, of obsolescence. The MAPI formula builds such projections into its formula and its charts.
- The formula should provide a means for comparing investment now with investment at some future time, thus taking possible deferment into account.
- The formula should make automatic corrections for taxes paid out of earnings arising from the investment.
- The formula should concern itself with loss in salvage value on the old equipment and capital consumption of the new equipment, which requires an estimate of the salvage value of the new equipment at the end of its service life.

An example, taken from MAPI files, is illustrated in Figures 6.1, 6.2, and 6.3. The calculated "urgency rating" is in the nature of a rate of return, but is not to be confused with other more common methods of calculation. This urgency rating may then be compared to similar ratings on other investment opportunities, selecting those investments with the highest ratings. It would be unwise, of course, to select any investments with urgency ratings below the interest on debt and the cost of equity capital needed to finance the investment.

PROJECT NO. _____ 5 _____   SHEET 1

SUMMARY OF ANALYSIS
(SEE ACCOMPANYING WORK SHEETS FOR DETAIL)

I. REQUIRED INVESTMENT

1 INSTALLED COST OF PROJECT	$ 112,000	1
2 DISPOSAL VALUE OF ASSETS TO BE RETIRED BY PROJECT	$	2
3 CAPITAL ADDITIONS RQUIRED IN ABSENCE OF PROJECT	$ 48,000	3
4 INVESTMENT RELEASED OR AVOIDED BY PROJECT (2 + 3)	$ 48,000	4
5 NET INVESTMENT REQUIRED (1 - 4)	$ 64,000	5

II. NEXT-YEAR ADVANTAGE FROM PROJECT

A. OPERATING ADVANTAGE
(USE FIRST YEAR OF PROJECT OPERATION)*

6 ASSUMED OPERATING RATE OF PROJECT (HOURS PER YEAR)		2,000	6

EFFECT OF PROJECT ON REVENUE	Increase	Decrease	
7 FROM CHANGE IN QUALITY OF PRODUCTS	$	$	7
8 FROM CHANGE IN VOLUME OF OUTPUT			8
9 TOTAL	$ A	$ B	9

EFFECT OF PROJECT ON OPERATING COSTS			
10 DIRECT LABOR	$	$ 1,280	10
11 INDIRECT LABOR			11
12 FRINGE BENEFITS		155	12
13 MAINTENANCE		3,500	13
14 TOOLING		2,000	14
15 SUPPLIES			15
16 SCRAP AND REWORK			16
17 DOWN TIME		1,675	17
18 POWER			18
19 FLOOR SPACE		500	19
20 PROPERTY TAXES AND INSURANCE	1,680		20
21 SUBCONTRACTING			21
22 INVENTORY			22
23 SAFETY			23
24 FLEXIBILITY		3,000	24
25 OTHER			25
26 TOTAL	$ 1,600 A	$ 12,110 B	26

27 NET INCREASE IN REVENUE (9A – 9B)	$	27
28 NET DECREASE IN OPERATING COST (26B – 26A)	$ 10,430	28
29 NEXT-YEAR OPERATING ADVANTAGE (27 + 28)	$ 10,430	29

B. NON-OPERATING ADVANTAGE
(USE ONLY IF THER EIS AN ENTRY IN LINE 4)*

30 NEXT-YEAR CAPITAL CONSUMPTION AVOIDED BY PROJECT:		30
A DECLINE OF DISPOSAL VALUE DURING THE YEAR	$	
B NEXT-YEAR ALLOCATION OF CAPITAL ADDITIONS	$ 10,000	
	TOTAL $ 10,000	

C. TOTAL ADVANTAGE

31 TOTAL NEXT-YEAR ADVANTAGE FROM PROJECT (29 + 30)	$ 20,430	31

*For projects with a significant break-in period, use performance after break-in.

■ **Figure 6.1** *Summary of MAPI analysis.*

157

III. COMPUTATION OF MAPI URGENCY RATING

32 TOTAL NEXT-YEAR ADVANTAGE AFTER INCOME TAX (31 – TAX) $ 10,215

33 MAPI CHART ALLOWANCE FOR PROJECT (TOTAL OF COLUMN F, BELOW) $ 2,576*

(ENTER DEPRECIABLE ASSETS ONLY)

Item or Group	Installed Cost of Item or Group A	Estimated Service Life (Years) B	Estimated Terminal Salvage (Percent of Cost) C	MAPI Chart Number D	Chart Percentage E	Chart Percentage x Cost (E x A) F
Gear Shapers	$112,000	20	20	1	2.3	$ 2,576
					TOTAL	$ 2,576

34 AMOUNT AVAILABLE FOR RETURN ON INVESTMENT (32 – 33) $ 7,639

35 MAPI URGENCY RATING (34 + 5) * 100   %   12

* Since the chart allowance does not cover future capital additions to project assets, add an annual proration of such additions, if any, to the figure in Line 33.

■ **Figure 6.2** *Computation of MAPI urgency rating.*

The proposal suggests replacing eight No. 6 Fellows Gear Shapers with four new No. 36 Fellows Gear Shapers at an installed cost of $112,000. The present machines are between 35 and 42 years old. They represent a liability, for major breakdowns could occur at any time. Major repairs are estimated at $6,000 per machine. This figure includes a charge of $1,000 for downtime attributable to the capital additions. Once they have been rebuilt, there is still no assurance that future capital additions will not be necessary in the future. Accordingly, the total rebuild cost of $48,000 is prorated over five years, with an annual charge of $10,000.

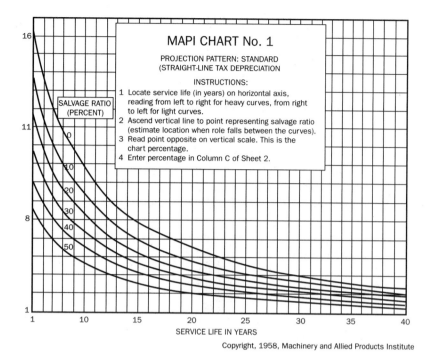

Figure 6.3 *MAPI Chart No. 1.*

The requirements of the work presently being handled on the No. 6 gear shapers are such that little precision is necessary. If it were, the present equipment would be unable to handle the work because of its worn condition. Although the ability of the proposed equipment to produce to closer tolerances is of little value for the present work, it will increase high-precision capacity. With the new units, there will be seven machines of this type in the department. This will make it possible to interchange work between the seven machines, whereas the present No. 6 machines are unable to handle some of the larger or more precise work. A value of $3,000 a year has been assigned to this greater flexibility.

A savings in direct labor will be realized from the ability of the new machines to produce approximately twice as fast as the old machines. This savings is estimated at $1,280 a year, plus fringe benefits of $155, or $1,435. In addition, their heavier and more rigid construction should produce greater cutter life. A total of $2,000 has been assigned to this factor.

It is estimated that approximately 300 man-hours of maintenance labor will be spent on the rebuilt gear shapers in the coming year. To arrive at a more realistic figure for the cost of one hour of direct labor spent in repairing a machine, the analyst attempts to account for the expenses involved in equipping and maintaining the repairman, as well as for the direct labor time. Accordingly, the figure of $10 an hour is used to reflect the full cost of the repairman. This is arrived at by establishing a burden rate for the Machine Repair Department in much the same way as such rates are calculated for a Production Department. Therefore, the cost of maintaining the old equipment is $3,000 a year (300 man-hours at $10) plus $500 for maintenance materials consumed. Maintenance on the proposed new equipment is assumed to be negligible.

Downtime can be expected on the rebuilt equipment, arising from ordinary maintenance. Delays in production will be inevitable, and expense will arise from the resulting bottlenecks. The total annual cost of this downtime is placed at $1,675. A savings of floor space of 250 square feet is anticipated, valued at $2.00 a foot for $500 annually. On the other hand, the new equipment is charged with $1,680 a year for additional taxes and insurance.

Assumptions

- 2,000 hours per year of operation.
- 20-year service life.
- Terminal salvage value of 20 percent.
- Tax rate of 50 percent.

References

1. Lindbom TH. Robots: Capabilities and Justifications. *Manufacturing Engineering & Management.* July 1972:17–19.
2. Helfert EA. *Techniques of Financial Analysis.* Homewood, IL: R.D. Irwin, Inc.; Revised 1967:154.
3. Terborgh G. *Dynamic Equipment Policy.* New York: McGraw-Hill Book Co., Inc.; 1949:143.
4. Terborgh G. *Business Investment Policy: A MAPI Study and Manual.* Washington, D.C.: Machinery and Allied Products Institute; 1958:168–171.

Effects of Legislation

7

The Williams-Steiger Occupational Safety and Health Act of 1970, which became effective on August 27, 1971, contains several sections that could affect the desirability of using industrial robots in industry.[1] Where costs to comply with provisions of the Act exceed the costs of an industrial robot or where costs to comply coupled with labor costs exceed this limit, then the industrial robot may offer a desirable solution. Each subpart is reviewed here as it might relate to robot applications:

1. Subpart D, Walking-Working Surfaces

   Where scaffolding and guarding of ladders and stairs are costly, robots may offer an economic solution to working in elevated locations.

2. Subpart E, Means of Egress

   In very tight and confined places, an industrial robot may perform an operation that would be impossible or not permitted by these new regulations.

3. Subpart G, Occupational Health and Environment Control

   This section sets standards for air contaminants, ventilation, noise exposure, and ionizing and nonionizing radiation. To comply with any or all of these requirements could entail considerable cost. Since industrial robots, like other forms of automation, are not affected by these environmental factors to any great extent, they may be considered for some of the hazardous or irksome tasks where these conditions exist.

4. Subpart H, Hazardous Materials

Handling of explosives, noxious chemicals, flammable and combustible liquid, toxic gases, and other hazardous materials offer opportunities for industrial robots where life and serious injury possibilities are involved.

5. Subpart O, Machinery and Machine Guarding

The new requirements for guarding machinery and offering operator protection devices make this a fertile field for industrial robots. This section specifically discusses the requirements for wood-working machines, cooperage machinery, abrasive wheel machinery, mills and calenders in the rubber and plastics industries, mechanical power presses, forging machines, and mechanical power transmission apparatus. Press feeding and unloading, and handling of hot parts for forging seems to offer good opportunities for robot application.

6. Subpart P, Hand and Portable Powered Tools

Operation of drills, riveters, saws, paint spray guns, and other similar hand tools may be handled by an industrial robot in a complex pattern on a repetitive basis. The ease with which a robot may be programmed or reprogrammed may offer economical solutions to these tasks.

7. Subpart Q, Welding, Cutting, and Brazing

Operator protection from sparks, molten, or hot metal and eye protection from intense light sources is discussed in this section. One of the largest robot installations in the world at the General Motors Vega Plant in Lordstown, Ohio, uses some twenty-eight industrial robots to weld car bodies together on the Vega assembly line.

8. Subpart R, Special Industries

Special applications and worker protection is described for the pulp and paper industry, the textile industry, bakery equipment, laundry machinery, sawmills, logging, and agricultural operations. Although few robot applications are known in these industries, special hazards may offer good opportunities for their use.

9. Subpart S, Electrical

Hazardous locations and high voltage are the primary areas where robots may be applied in this area.

A copy of a more complete treatment of the Williams-Steiger Occupational Safety and Health Act of 1970 is found in Appendix A taken from the Federal Register, Vol. 36, No. 105, Saturday, May 29, 1971. The Walsh-Healy Public Contracts Act also lists limits to occupational noise exposure, and its regulations are produced by Williams-Steiger Occupational Safety and Health Act of 1970.[2]

References

1. *Federal Register*. Vol. 36, No. 105, Saturday, May 29, 1971. Title 29—Labor, Chapter XVII, Part 1910, Occupational Safety and Health Standards. 10466–10645.
2. What Do You Know About The Walsh-Healy?, *National Safety News*. September 1969:48–51.

8

Human Factors

The image of the automatic factory filled with industrial robots replacing all human hands is rapidly being dispelled. A spokesperson for the United Auto Workers has stated: "In regard to industrial robots, our policy approves of them—as long as they are introduced in a way to accommodate the needs of human beings who are affected."[1] Robots then are becoming co-workers with labor, to assume some of the more dangerous, tiring, or otherwise irksome tasks and to free labor for the more creative or more adaptive work where greater job satisfaction may be realized.

The burgeoning technology of automation will certainly dictate the character and contours of the industrial environment of tomorrow's blue-collar worker. The currently available data suggest that the advent of automation will have the following short-range effects on the blue-collarites' world of work:

1. **Improved physical working conditions,** especially in regard to bodily exertion, health, safety, and nervous strain due to excessive sound levels reduced.

2. **Increased emotional problems,** such as gastric ulcers, resulting from isolation from other workers and prevention of daydreaming by the need for increased, constant attention. These effects may be temporary or characteristic only of transitional phases of automation; they have sometimes been mitigated by training programs.

3. **Little upgrading of workers,** contrary to popular opinion and hope. Skill levels seem to decline more than they rise. In the case studies undertaken by the Bureau of Labor Statistics in a variety of representative industries, the general effect of automation was to substitute one low-skilled job for another.[2]

4. **Decreased control of work pace** and less handling of materials. A representative automobile worker told one interviewer: "(I don't like) the lack of feeling responsible for your work. The feeling that you're turning out more work but know it's not yours really and not as good as you could make it if you had control of the machine like before."[3]

There is, however, a promising pair of developments that could alter the role of the blue-collar worker; at one significant point, the emerging ideas about work and the emerging technology intersect. It is here that we find the most exciting possibilities regarding the future of toil. Georges Friedmann, in *The Anatomy of Work*,[4] reviews experiments in Europe and the United States since World War II—experiments that endeavor to overcome the "orthodoxy of scientific management," which decrees that maximum efficiency demands specialization and routinization of the work process. Such firms as IBM, Detroit Edison, and Equitable Insurance have reversed the division of labor through "job enlargement," giving each worker a whole sequence of tasks constituting a complete job and providing some sense of achievement, or letting workers handle their jobs in small groups, rotating among different tasks and setting their own working conditions. The results have been good for the workers and good for business. Although the workers do not perform their more complex and varied task with the same robot-like efficiency, the reduction in absenteeism, botched work, and employee turnover has usually more than made up for the more relaxed pace of work.[5]

Taken in the context that robots are merely a subclass of automation, the most promising area of application lies not in merely displacing the worker on the assembly line, but in becoming a tool to be used to free the worker to more varied and creative tasks and a greater sense of accomplishment.

References

1. Connole AW. The Human Aspects of Automation. *Proceedings of the First National Symposium on Industrial Robots*. Chicago: IIT Research Institute; April 1970:96.
2. Buckingham W. *Automation: Its Impact on Business and People*. New York: Harper & Row, Pub.; 1961:98.
3. Ibid., 93–108.

4. Georges Friedmann. *The Anatomy of Work*, translated by Wyatt Rawson. New York: The Free Press of Glencoe, Inc.; 1961:120.

5. U.S. Department of Labor. *Manpower Implications of Automation.* Washington, D.C.: Government Printing Office; 1965:32, 37, 39, 43–49.

167

9

Safety Standards for Industrial Robotics

Industrial robots use is increasing, and so the need to understand robot safety issues affects an ever-growing cross-section of industries, applications, and personnel. Today, a variety of safeguarding devices and resources are specifically designed for robot systems. Successfully integrating these devices into a work cell requires a solid grounding in fundamental engineering, a dose of insight into human behavior, and a good understanding of industry standards and compliance issues.

Safety affects everyone, and safe robotic systems are key to a safe workplace. To learn the latest tips and insights into robotic safety, most people turn to the Robotic Industries Association (RIA), the official trade association that sponsors the American National Standard for Robot Safety, as well as the Annual National Robot Safety Conference.

The Robotic Industries Association (RIA)

Founded in 1974, RIA is the only trade group in North America organized specifically to serve the robotics industry. RIA, an Ann Arbor, Michigan-based not-for-profit trade association, sponsors the National Robot Safety Conference, which features in-depth workshops, classroom sessions, expert panels, and a networking reception with tabletop exhibits from leading suppliers of robot systems and safety equipment. The RIA serves as the secretariat for the Robot Safety Standard.

The Association is also well known for its International Robots & Vision Show and Conference. RIA member companies include leading robot manufacturers, users, system integrators, component suppliers, research groups, and consulting firms. Case studies, technical papers, application information, news, editorial and event information, as well as an online Buyer's Guide and an interactive "Ask the Experts" forum are available on Robotics Online (www.roboticsonline.com).

Risk Assessment Software from RIA

The Robot Risk Assessment CD walks you through the stages of a risk assessment following the guidelines set in the ANSI/RIA R15.06 Robot Safety Standard. This easy-to-navigate program helps you conduct hazard analysis based on the application and the associated tasks the robot will be completing. It then leads you through the risk assessment where you will determine the severity of potential injuries and the frequency of exposure and likelihood of avoidance. Once the risk assessment is complete, the program provides for easy documentation.

RIA is working with the International Standards Organization (ISO) on the international robot safety standard (ISO 10218) and other safety issues to ensure up-to-date safety standards for the robot industry. Customized in-house training is available from the Association, and regional and national robot safety events are held each year to keep up with the changing marketplace.

National Robot Safety Conference Presentations

What do RIA members have in store when attending a National Robot Safety Conference? Several people from RIA companies plan presentations in the form of workshops and case studies. These provide attendees with the opportunity to learn and sometimes "experience" the situation first-hand from professionals who deal with such problems on a day-to-day basis.

Safety Circuit Design

One presenter ran a workshop on safety circuit design. A variety of examples of safety circuitry were shown to demonstrate designs that are in compliance with the R15.06-1999 standard. Discussions included fence circuits, interlock gate circuits, and circuits that use light curtains or floor mats to shut the robot down when someone is loading/unloading parts within its reach.

That workshop included design elements for muting circuits and zone switches. These allow the robot to operate on one side of the cell while the operator is on the other side, while preventing the robot from moving to the side that the operator is on. Attendees were given warnings on certain things such as not taking a safeguard device and wiring it into the inputs of a PLC to program safety. This is not in compliance with R15.06-1999.

Safety Code Compliance

A Senior Staff Engineer at Underwriters Laboratories (UL), New York, also presented at the NRSC. His case study was based upon a client request to evaluate a robot work cell that will be integrated or a robotic system that is already installed. The group performed a review to determine if a work cell is in compliance with R15.06-1999.

Just like at the UL, attendees were prompted to look at fundamental safeguarding applications, such as interlocked gates, door switches, light curtains, and laser scanners. This helps to make the customer and integrator more familiar with R15.06-1999 guidelines and bring the work cell into compliance.

New Technologies

A John Deere employee presented case studies on new technology standards. He talked about dynamic limiting, which is based used when the system integrator wants the operator to be in the hazard area. Dynamic limiting is a strategy of redefining safeguarded space by using electronic devices. This requires the removal of motor power at some times, but the use of motor power when needed.

Another case study covered remote axis control. Such issues as slow speed requirements on existing cells and older cells were discussed. The employee also talked about hard stops. It was recommended that attendees put in hard stops to prevent a robot from moving any further than is actually needed. In addition, the presenter evaluated discoveries from risk assessment, which he referred to as a "Wall of Shame."

"The Wall of Shame displays blatant safety violations. I will show photos of hazards we found through risk assessment. We do a quick risk assessment and show how there are a lot of simple things that can be done incrementally that will get you a lot of value," the presenter noted.

Human Factors

Yet another presentation was entitled "Ergonomics and Automation: Human Factor Considerations in Machine Controls and Designs." The subtitle was "How to Avoid Pitfalls at Your Factory."

Questions and Answers

Roberta Nelson Shea, chair of the RIA's 15.06 Safety Standard Committee for several years, presented generally asked questions about the R15.06-1999 safety standard. She also gave a workshop on safeguarding device selection that will go through various technologies, how they are applied, and when or where someone can use them. "Applying a safeguarding device comes down to how much access is needed. If minimal, an interlock barrier is sufficient. If access is frequent, use a presence-sensing device," said Nelson Shea.

Other questions Nelson Shea said need asking included, "Are there projectiles?" "If so, and you need frequent and easy access, integrators would have to modify the work cell. The need to safeguard against projectiles is not only for the perimeter of the work cell, but also of each task that is planned or envisioned for the system itself." Nelson Shea advocates interlock barriers to contain objects that could be thrown.

Jeff Fryman, Director of Standards of the RIA said this about one of *his* presentations: "I will present an in-depth review of the R15.06-1999 standard and help people understand it. People who circumvent the standard or ignore it, do so at their own peril."

172

The Robot Safety Standard

The Robotics Industries Association has been busy "getting out the word" that the robot safety standard, ANSI/RIA R15.06-1999, has been approved and published. They continue to see the standard embraced by an ever-growing group of robot users.

The ANSI/RIA R15.06-1999 Robot Safety Standard is addressed thoroughly in workshops, sessions and panel discussions. The RIA actively participates in companion standards development activities including ISO and other machinery safety efforts. There are also sessions on the Canadian Safety Standard, CSA Z434, Ontario's pre-start health and safety reviews, and developments in new technology for control reliable circuits.

"Robots need to be safeguarded like any other machine," said Jeff Fryman, Director of Standards, RIA. "We have a standard for this purpose, and we provide in-depth training on the standard as well as risk assessment to help a company ensure their workers are properly protected and to improve their compliance with OSHA requirements. In fact, we have released an updated version 2.0 of our Risk

Assessment Software that will be available as a computer-based training workshop."

R15.06-1999 Safety Standard

The robotic safety standard, ANSI/RIA R15.06-1999, provides procedures for industrial robot manufacture, remanufacture and rebuild as well as robotic system integration. The standard is a set of voluntary guidelines of methods for safeguarding operators.

An important feature of R15.06-1999 is assessment of a work cell and determination of potential hazards that an operator could face. The first step in laying out a work cell is to design-in safety. Once designed, the work cell goes through a risk assessment of potential hazards.

RIA's own Jeff Fryman summarized the R15.06-1999 and the risk-assessment process. "R15.06-1999 is broad-based and makes robot makers, integrators and operators aware of their responsibility to safeguard people in or near a work cell. Investing in safety is a cost-avoidance issue," Fryman said. "Those doing risk assessment had their eyes opened going through the process. Some not-so-obvious safety issues were taken for granted until integrators, end-users, and robot makers did risk assessments." Fryman is RIA's Director of Standards Development and was key in formulating R15.06-1999.

Wayne Maynard, Director of Ergonomics at the Liberty Mutual Insurance Company, Hopkinton, Massachusetts, also had a major hand in coming up with R15.06-1999's guidelines. "R15.06-1999 offers very good guidelines on risk assessment for robotic operations. This is a great enhancement over previous versions of the standard," Maynard said. "I especially like the guideline on how to implement risk assessments. This guideline was created by people in the robotics industry, not just safety professionals. For this reason, R15.06-1999 has a great deal of credibility."

Because R15.06-1999 is a guideline and not a fixed regulation, there is inherent flexibility in how to comply. "There are many ways of keeping people safe, and there is not just one style for every situation," Gil Dominguez said. Dominguez is Senior Engineer at Rimrock Automation, Inc., New Berlin, Wisconsin. The need for flexibility is recognized and built into the R15.06-1999 safety standard.

Safe work cell design, although important, cannot prevent every hazard. After design, installing the appropriate safeguarding device is the next order of business. Liberty Mutual's Wayne Maynard speaks to that.

"The first step is to design-out hazards from happening. If this can not be done, the best course of action to take is to install the proper safety system to shut down the work cell if someone enters a dangerous area." An ideally set-up safeguarding device would not allow the robot to restart until the person leaves the hazardous area. Maynard concluded by saying "R15.06-1999 addresses shut-downs nicely, especially with E-stops." R15.06-1999 provides guidelines to prevent people from entering an area that is not safeguarded.

Current Issues That Affect the Old Standard

As the Robotic Industries Association "point man" (officially the Director of Standards Development) for robot safety, Jeff Fryman has had the pleasure of meeting and working with a number of robot users in plants. In many cases, this has been as the instructor conducting the Robot Safety Standard (ANSI/RIA R15.06-1999) course at an in-house training seminar. This is just one of several courses the RIA offers as part of its industrial robot safety training program. In-house training seminars are one of a variety of robot safety resources the RIA offers to the industry.

Starting in 1999, these courses were intended to introduce the "new" robot safety standard to the people who needed to know about it and comply with its provisions. The word "new" is qualified because this version replaced the 1992 edition of the standard and contained new and important safeguarding strategies and guidelines.

One-on-One with the Director of Standards Development

Jeff Fryman often answers questions about the robot safety standard. Following is an excerpt from an interview with him regarding the standard:

One question that is repeatedly asked is, "Can I use my old robot in a new work cell?" The answer is a definite yes, with a number of caveats.

First, as background I would like to comment on the thoughts of the committee as they drafted the current standard. I believe the committee was very sensitive to the needs of industry, including consideration of cost issues. They acknowledged these cost impacts by ensuring that there was a valid return on improved safety for the investment cost in implementing new requirements. Safety has a cost, but a sound safety program would look like a good investment compared to the costs of litigating a serious accident.

"Grandfathering," or the requirement for retrofit, was the single most contentious issue in reaching consensus on the standard. One camp did not want to look back, the other camp felt noncompliant installations were not entitled to another "free ride" for having done something wrong or inconsistent with the previous standard. Both camps agreed that the provisions of the '99 standard provided improved safety and should be required for all future installations.

How did this play out? Basically the committee looked at two aspects of a robot system—the robot itself, and the work cell.

The robot hardware itself is "grandfathered" from retrofit (though voluntary upgrades are allowed). This was a concession in consideration of the fact that hardware is difficult to change, and may not be cost effective. An example of this would be the provisions for limiting devices, particularly hard-stops, on some of the axes. Though not difficult to include when designing the robot, installing hard-stops could be difficult to impossible on existing hardware.

On the other hand, the system or work cell was easier to change, and over time typically is changed and upgraded in the normal course of business. Thus, safety enhancements could be made incrementally and their costs amortized over the useful life of the robot application. Typically a robot work cell would have a useful life on the order of 5–7 years before its technology (albeit not the hardware) was obsolete. Cost depreciation and amortization would allow for an orderly and economical replacement for older systems. Cost trade-offs for new, remanufactured, or rebuilt robots would be part of the business case made when new systems are designed.

How is this implemented in the standard?

Clause 1.3.1 New or remanufactured robots—provided for a grace period of 24 months intended to allow the manufacturers time to re-engineer their products and clean out existing inventory of the older non-compliant models. Remanufacture—the changing of the capabilities or functionality of the robot with new and upgraded capabilities; i.e. new controllers, stronger servo actuators and drive train to increase payload, etc., including re-engineering by a third party—is differentiated from rebuilding—the complete overhaul and refurbishment to original specifications, albeit with newer components and latest software releases. The business decision to invest in remanufacturing a robot versus rebuilding a robot includes the decision to accommodate updating the robot to the new safety requirements in the standard.

Clause 1.3.2 A rebuilt or redeployed robot is the final part of the robot hardware equation; the robot hardware is "grandfathered" to the requirements of the standard effective on the date of manufacture. Safety enhancements are allowed, and encouraged, but it can be an "a la carte" selection—those features that are easy or economical to do rather than the whole list.

Redeployed robots can be redeployed from one cell to another cell in a similar application or can be redeployed to whole new applications where only the end-effector (complete with supporting hardware) is changed, and the task program is changed. Now this is what is allowable, but not necessarily feasible given the specialization of some robots. For example, a robot can go from one MIG welding application to another MIG welding application but on a different part. A robot can also go from a welding assignment to a material handling application by changing the end-effector and the task program. This may be useful in utilizing older robots that no longer have the repeatability tolerances necessary for precision work. An old robot with a repeatability error of 2–3 mm may work just fine in a palletizing application where the carton stack would not notice a 3-mm error, compared to a welding application that could result in a total bead void in the same circumstances.

So, now the conundrum. Since new work cells must be in compliance with the new standard, using older, "non-compliant" robots in new cells may require some compensation in the system installation to provide the appropriate level of safety. Again, depending on the age/design of the robot, some of the additional safeguarding requirements may not be practical, even though it is allowed for in theory. These are all considerations that must be looked at in the business plan to decide on the design of new work cells and the components that go into them.

The biggest, but not the only issue may be the necessity for a dual run chain to meet the requirement for "control reliable" circuitry. Most older robots only had a single run chain. The good news is an external run chain with monitoring, and an additional set of motor contactors can be added to satisfy this requirement. Of course, this may be much easier said than done. Control reliability will be needed if you follow the prescribed methodology in Clause 8, and may be required based on your determination of risks from a risk assessment following the procedures in Clause 9.

Another issue could be the lack of limiting devices to establish the required restricted space. This may necessitate a larger footprint for the work cell to achieve the necessary clearance, or applying additional safeguarding is required.

Selection of certain optional operational features may also require accommodation. Using a teach pendant inside the safeguarded space requires an enabling device, but not necessarily the three-position enhanced enabling device. The use of high-speed attended program verification will require an entire list of specific compliance items including the enabling device, control reliable circuitry, and adequate clearance.

I try to answer these and many other questions in our in-house seminar offerings, and in our annual robot safety programs, the spring regional and the fall National Robot Safety Conference. See our Website for full details on the safety conference and the other robot safety resources available to you through the RIA (www.roboticsonline.com).

Safety Standard FAQs

(Questions were provided by Joe Campell, VP of Marketing for Adept Technology, and answers provided by Jeff Fryman, Director of Standards for the RIA.)

Q. I keep hearing about the RIA standard. Is it going to be law? As a system integrator, am I legally bound to do something different in my designs when this becomes effective?

A. The "RIA standard" is ANSI/RIA R15.06-1999, Safety Requirements for Industrial Robots and Robot Systems. It is not a law; it is a voluntary American National Standard. Hopefully your system designs already comply, but again it is not a legal issue so much as a compliance issue. As more and more users require compliance with the standard in their contract specifications, you will be obliged to comply with the standard. And more contracts will be let this way, since the user is required to comply with OSHA directives that include voluntary standards by reference.

Q. I'm a machine builder, and I'm just going to tell my vendors that whatever I buy from them has to meet the new standard. That's enough, isn't it?

A. The standard is targeted at personnel safety, and does not provide any detailed hardware design information, only performance criteria. If your vendors do the engineering, then telling them to comply with the standard may be sufficient. However, if you are a machine builder, as stated, then you probably do the engineering and will have to comply with the performance requirements of the standard.

Q. CE? RIA? ANSI? Is there one standard I can design to that will meet all needs?

A. Simply stated, no. CE refers to European conformity and has nothing to do with suitability for use in the United States. ANSI is the accrediting agency that approves the standards sponsored and written by the RIA, a Standards Developing Organization. The R15.06 is performance requirements for personnel safety. A design standard is ANSI/UL 1740 that actually states hardware requirements and specifications. The UL 1740 and R15.06 are harmonized so that if you build the hardware in compliance with UL 1740, you should be able to meet the safeguarding requirements in R15.06.

Q. I have an old robot that's still working fine, but the vendor can't support the controller and I want to upgrade it. Do I have to worry about the RIA standards?

A. Worry? No; Comply? Yes. The upgrading of a robot with a new controller would be considered "remanufacture." As such, the remanufactured robot will have to comply with all the performance requirements of the 1999 R15.06.

Q. I'm a robot manufacturer, and I sell a lot of used and remanufactured robots, with their original controllers. Is this going to be a problem when the RIA standards take effect?

A. No, this should not be a problem. However, where you used the word "remanufactured" I think you meant, "rebuilt" in the new standard criteria. Rebuilt robots retain the same configuration and capabilities of the original robot and only have to comply with the standards in effect on the original manufacture date. If you actually "remanufacture," i.e. upgrade the robot, then you will have to comply with the 1999 edition of the standard.

Q. I'm a used robot vendor. Can I keep selling used systems after the RIA standards take effect?

A. Yes, certainly. However, you need to review for the compliance criteria if you do any work on it prior to selling it as used equipment.

Q. I'm a robot vendor. Can I ship used robots to Germany today?

A. Yes. However, if you want them to be used as something more than a boat anchor, you need to do your homework for an EC declaration of conformity prior to it arriving in the European port of entry. If you are not the manufacturer of the equipment, compiling the technical construction file may prove very difficult. You will have to comply with the implementing standards for the machinery directive as well as the low voltage directive and the EMC directive. Oh, and you will have to have the machine instructions available in German!

Q. I keep getting sales calls from consulting companies offering to help (for $$$) me meet the safety regulations. What's the role of UL and TÜV and all the other consultants? Did they write the regulations? Who authorizes or certifies them to interpret the laws?

A. A number of very reputable consultants are available to help with standards compliance if you feel uncomfortable with interpreting them or just want a helping hand. These consultants may be independent, or work for one of the third-party certification organizations such as UL and TÜV, among others. Some of the consultants and representatives from the organizations did sit on the drafting committee, so they are well versed on the intent of the standard. UL also offers a field certification program for robot installations; of course they look at UL 1740 as well as R15.06. There is no certification program, so in our free enterprise system, you do need to adopt the "buyer beware" attitude when selecting your consultant.

Q. I hired a consultant to review the systems I build. Does that guarantee it's OK to ship to Europe?

A. Nope, no guarantee. Only a signed declaration of conformity certificate by a representative of your company will assure entry of a machine into the EU.

Q. I'm losing business to a competitor who isn't meeting the standards with their products, but they say they are. What can I do?

A. This is a continuing concern to us because there are so many good companies out there working hard to comply with applicable standards. There is no conformity compliance program in the United States, so there is no independent arbiter to determine if a product complies or not. Voluntary third-party certification can demonstrate this, but again is only useful to the company complying. It does not prove something doesn't comply. An educated consumer is the best defense. The RIA is working hard on consumer education, conducting two robot safety programs each year, in addition to in-house training seminars, providing speakers for other forums, sponsoring articles, and a Website.

Safety First: A Review of Robotic Safeguarding Devices and Issues

Many factors need to be considered when choosing any device for use in a work cell, including plug-and-play considerations, conclusions to be made from risk assessment, and so on. New industrial robot technology and safeguarding devices have flourished for decades in North America as companies seek higher production yields and safer work environments.

"The overriding interest of a manufacturer is to keep things running. Whenever a safeguarding device signals a robot to stop, it should be because of a personal safety issue and not a nuisance stop nor because the unit is unreliable," asserts Roberta Nelson Shea of Honeywell Sensing and Control. Nelson Shea is the Safety Business Development Manager at the Freeport, Illinois, maker of robotic safety equipment. She continued by saying, "The biggest challenge is to try to integrate safety equipment and production equipment to have maximum up-time yet to still ensure that people are safeguarded."

If a safeguarding device impedes an operator's need for cell access, or is not intuitive, it might lead to unsafe behavior by manufacturing personnel and maintenance workers. For example, a worker might trick or bypass a safety device to enter a cell to clear a jam, a dropped part, or something else. These are known as *jumpers*, people who jump past the safety devices for the sake of production or maintenance issues. When this happens, the chances increase greatly for personnel to be injured when the safeguarding device fails to halt the robot for a legitimate safety concern. The optimal way to prevent this nightmare scenario is to use safeguarding technology that is robust and proven reliable.

Of course, budgets and physical connectivity are other things to keep an eye on when selecting and integrating a safeguarding device. "One problem is how to physically connect the safeguarding device to the robot," says Thomas E. Tabaska, Quality Manager at John Deere Worldwide, Horicon, Wisconsin. "You can spend lots of money on a reliable safety control device, but if it uses only one wire or one set of contacts to connect it to your system, the device is only as robust as that connection."

Tabaska spoke of other potential hurdles that integrators must over-come when linking safeguarding devices with a robot. "Integrators have to think about the strategy of when and how they want the safe-guarding device to engage and when they do not want it to." For exam-ple, if the robot is not at a particular station at a particular time, the safeguarding device need not be protecting the operator there. The safe-guarding device can disengage. This is called "muting" the safeguard.

Tabaska makes another important point that integrators and operators need to remember. "It is not necessarily the robot that is the danger. The hazard could be moving tables, parts, servo motors that turn or move a part, or extra axes that perform part loading." Operators need to be safeguarded from these systems as well as from the robot itself.

"Determining the type of safeguarding device to be used in a work cell begins with the design process," says Steven Freedman. "When a cell is designed, hopefully safety is taken into account to minimize potential hazards." Freedman is the Director of Safety Systems at SICK, Inc., Minneapolis, Minnesota, a maker of robotic safety devices. While emphasizing the necessity for safety, Freedman added that the solution has to be cost-efficient in order to be effective.

Some higher technology solutions to safeguarding human operators in a work cell include light curtains, laser sensors, and presence-sensing devices. The purpose of all these devices is to shut down the robot com-pletely, or halt the movement of the robotic arm, if a person is in an area within a robot's work envelope.

In hazardous applications, such as working with explosive or radioac-tive materials, the robot itself is the safety device. "Radioactive or explosive materials are the reason that a robot is used. A robot is used not to increase productivity as in industry, but to remote people from the hazardous operation," said Dr. William Drotning. Dr. Drotning is Project Leader at the Robotics Center in Sandia National Laboratories, Albuquerque, New Mexico.

Robot Safety Begins with the Design Process

Stacy Kelly, Safety Systems Product Manager of SICK, Inc. answers many questions about starting out with a safe robot system: "To realize the many benefits that robots offer a production facility, safety considerations are *the* top priority in protecting the operator, maintenance personnel, and other personnel that interact with the robot. These safeguards should be designed into and around the robotic cell early in the design process to maximize the inherent safety of the overall system. Beyond the safeguarding products, the planning and implementation of these products are of critical importance for good safeguarding practices and factor heavily into optimizing system design and cost. What are the potential hazards of the robotic cell? What safeguarding technologies are available? How do I keep out unnecessary personnel, yet protect necessary personnel? How much panel space must be used for relays? How difficult or easy will the troubleshooting of the system be? And, of course, what is the overall reliability and safety of the system?"

Risk Assessment

The first step in designing a safe robot system is to understand the hazards that exist in the system. This is commonly achieved through a formal risk assessment process that identifies and documents all production and non-production tasks and the hazards associated with them. The hazards are then classified based on the criteria, such as the severity, the potential injury, frequency of access to the hazard, and the possibility of avoidance. Risk assessments should be performed during the design phase and prior to the commissioning to ensure that no new hazards have arisen in the integration process.

Safeguarding Technology—Availability and Implementation

System designers must understand the current safeguarding technology and how this technology will save them time and money, both now and in the future. For example, the capabilities and size of the safety relays will dictate the amount of panel space that will be needed for wiring and relays. Availability of safety bus technologies will also have serious impact on the design and implementation. Optimizing this system will also provide flexibility for future expansion and minimize troubleshooting.

System designers must then know how to properly apply the safeguards. For instance, designers must understand safety distance, or the distance from a hazard the safeguard must be mounted to ensure the hazard will cease to exist before personnel can reach it. These calculations should be done at the initial commissioning of the system and periodically thereafter to ensure that the safety distance has not changed due to mechanical wear or other changes in the system.

Perimeter Guarding

To keep unnecessary personnel out of the restricted space of a robot cell, one of two safeguarding methods are often used. Hard-guarding is a fencing-type solution. With hard-guards, door access will likely be needed, and these doors must contain interlocking devices (e.g. safety interlocks) to guarantee safe access. Optical perimeter guards (e.g. light curtains) are a more flexible solution that can easily adapt to layout changes. These guards must be located at a safe distance from the hazard and must interface with the robotic control system. Optical perimeter guards are often used in combination with hard-guards. An additional requirement for either perimeter solution is that the operator control, such as a system reset, be located outside the safeguarded area.

Protection on the Inside

If there is a danger to the operator, maintenance personnel or other personnel from robotic motion within the restricted or operating space, this area must also be safeguarded. Area safety scanners are often used in these areas, as the scanner coverage area is wider and more flexibly programmed than with other devices. Light curtains have also been used. Again, these safeguarding devices must be located at a distance that provides adequate stopping time of the system and accounts for the speed of approach from the personnel in the area as well as a depth penetration factor, as defined in the ANSI/RIA R15.06-1999 standard.

The Solution

How should a company ensure that it has a reliable and safe robot system? Good planning, proper installation, and ongoing support are paramount. Some companies may decide to develop this expertise internally, while others turn to outside safety consultants and manufacturers of safeguarding systems. Safety consultants and manufacturers would certainly be in a better position to keep up with the rapidly changing technologies and industry standards. Either way, the goals of a safe and productive environment are better achieved when considered together.

Designing a Safe and Highly Productive System

Steve Freedman, Safety Division Manager at SICK, Inc., elaborated on how companies can ensure they have a safe system by designing properly from the onset of a project. "Start thinking safety early. One of the most frequent mistakes system designers make is thinking only of the production process and ignoring the fact that the system will be safeguarded later to prevent injury before it is put into productive use. The result is that operators often find the safeguards inhibit their ability to perform their jobs efficiently, resulting in reduced productivity.

Many operators unwisely bypass the safeguards in order to keep a sustained production pace or troubleshoot equipment. This has become one the most frequent causes of machine related accidents in automated production today. Training personnel is critical to avoiding this. However the safest systems are those where the required safeguards are considered simultaneously with the production system design. In these cases, systems are designed that allow personnel to safely and easily perform their jobs with no need to bypass the installed safeguards. As a side benefit, productivity can often be increased dramatically while helping to meet the goal of zero injuries.

Properly designing a safeguarding system is not a simple task. It first requires a clear understanding of the hazards that exist in the system. This is commonly achieved through a formal risk assessment process that identifies and documents all production and non-production tasks and the hazards associated with them. The hazards are then classified based on criteria such as severity of the potential injury, frequency of access to the hazard, and the possibility of avoidance. Risk assessments should be performed during the design phase and prior to commissioning to ensure that no new hazards have arisen in the integration process.

Once the hazards have been identified and classified, up-to-date knowledge of current safeguarding technologies is crucial. There have been many advances in safeguarding technologies that allow for greater flexibility in the design and integration of the safety system. The designer should be well versed in the productivity enhancing side of safeguarding as well. Functionality such as *Presence Sensing Device Initiation or PSDI*, which allows the safeguard to perform double duty by safeguarding personnel and initiating the machine cycle, has significantly improved productivity for many companies while reducing the ergonomic repetitive stress of a separate cycle-start button. It is always advisable to select safety components and interfaces that have been third party certified to local standards by an accredited agency to ensure their intended functionality.

System designers must then know how to properly apply the safeguards. For instance, designers must understand *safety distance*, or the distance from a hazard that the safeguard must be mounted to ensure that a hazard will cease before personnel can reach it. Safety distance calculations should be made at the initial commissioning of the system and periodically thereafter to ensure that mechanical wear on the system over time has not caused the hazard's stopping time to increase thereby increasing the required safety distance.

Properly interfacing safeguards with the machine control and/or e-stop circuitry is another area that requires serious attention. Depending upon the risk assessment, safety circuits may be required to be designed such that component failures will not prevent hazardous motion from stopping, and that restart will be prevented until the fault is corrected. This can obviously force an extremely complex circuit design. Prepackaged safety relays (DIN Rail mountable pre-wired circuits) are a readily available way to reduce safety circuit design time and wiring errors. Safe PLC's and safe bus networks are also becoming popular ways to safely reduce safety wiring and troubleshooting costs for companies on the forefront of safety technology.

So how should a company handle these complex issues surrounding the safety of their personnel and ensure that their systems are designed properly? Many companies develop this expertise internally, while others turn to outside safety consultants and manufacturers of safeguarding systems for their expertise rather than try to keep up with the rapidly changing technologies and industry standards. Either way, the goals of a safe environment and a productive environment are better achieved when considered together.

How to Stay Competitive

As manufacturers continue to explore more exotic, lightweight materials in their quest to capture the consumer's imagination, robotic integrators and suppliers will follow suit, building better tooling and more efficient automation systems for a highly competitive industry.

RIA provides a number of resources for information about top supplier companies for tooling, welding services, new robots, and peripheral equipment. Request the free Robotics Industry Directory or visit the Buyers Guide on Robotics Online to search for more solutions, and learn more about successful applications in the "Tips & Tech Papers" section of Robotics Online.

Safety First, Last, and Always

When equipment design measures are combined with personnel safeguarding measures such as passive sensors that detect human presence in the robotic cell and new safety standards developed by the Robotic Industries Association (RIA) and its member companies, the result is a safer, more efficient workplace with less downtime.

Safety and downtime are closely linked. "A big issue with older robots was that they got lost when the cycle was interrupted, prompting workers to try and fix a problem without stopping the cell," said Jeff Fryman, director of standards development at RIA. "The robots had to be driven home and a lengthy restart process had to be started. A key element in safety is to remove the incentive for a worker to do something unsafe. We have to remove the incentive for the employee to bypass the designed safeguards and provide a robot system that doesn't have a lengthy restart process."

10

The Impact of Machine Vision on the Robotics Industry

Most people have little idea of the advances of machine vision, other than their observation of the wand that reads the bar code on the groceries at the check out counter. The manufacturing industry is required to have more flexible and reliable production problem solutions than even before. Lead times have shortened, part identification complexity has increased, tooling budgets have tightened, and zero defects are expected. In addition, increased liability and warranty risks require improved traceability. These are tough challenges, providing a level of confidence required by manufacturers. Second-generation low-cost vision sensors with networking capabilities are extending these applications to ever-widening audiences.

Machine Vision

The goal of most vision systems is to improve quality and productivity in the manufacturing process. In a typical application, a presence sensor detects a part and signals the vision system to activate a camera, positioned to capture an image of the part, and send it to an image processor. Machine vision is superior to manual inspection because it can handle high speed and volume production, offers more precise readings, doesn't get tired, and provides better process control. Unlike a human inspector, machine vision systems can not only report, document, and inspect for failures, but can also provide the reason for and the extent of that failure. Machine vision's ability to avoid false accepts and rejects is especially important to manufacturers who are under pressure to maximize throughput with minimal wastage.

At the present time, the majority of vision applications are for machine guidance or inspection. In quality control inspection, the vision system determines whether parts or subassemblies are acceptable or defective, and then directs sorting equipment to reject or accept them. Machine guidance vision systems are used to improve the accuracy and speed of assembly robots and automated material handling equipment.

Categories of Machine Vision

Code Reading

In the medical and pharmaceutical industries, there are mandates in place that insist on full traceability of products and devices, and now other industries, such as automotive, are making similar demands. To meet this requirement, companies are laser-etching Human Readable Alpha Numerics and 2D data matrix codes onto their products and using machine vision to read the codes. Manufacturers can then take advantage of the information on these codes for tracking, verification, and statistical quality control. It is interesting to note that, "If you used an old bar-code reader, it would need to be connected to a PC linked into the plant network. DVT smart image sensors have Ethernet on already, thereby removing the requirement for any PCs on the shop floor," said Philip Heil, Chief Technology Officer, DVT.

Print Verification

Machine vision techniques are being used to make sure parts, components, and products are labeled correctly. In the real world of the production line, this can be more complex than it initially might seem. The system may have to compensate for variations in character density, inking, and shape, as well as secondary effects of laser etching, stamping, and engraving. For just that purpose, the best machine vision technology allows users to adjust the application quickly, with little or no delay in the production process.

Robotic Guidance

Of all the forms of automation, robotics is among the most flexible. Machine vision enhances that flexibility by enabling the robot to see the part or subassembly on which it is working. Machine vision can provide real-time live feedback to guide the robot as it goes through its programmed sequence. To perform this level of guidance, a vision system usually locates parts for the robot to pick up, identifies the correct location and orientation at which to place the parts, and sends this information to the robot for execution. Speed and accuracy are essential for this process. One recent application of robot guidance required the machine vision system also to determine the spectral quality of lettuce as it guided a spider-like robot in picking ripe produce in the fields at Kalamazoo, Michigan.

Dimensional Gauging

For many products, quality is a matter of small dimensions and details. The latest machine vision systems excel in ensuring that those measurements are correct. These applications include part and hole positions, lines, angles, arcs, and diameters.

Flaw Detection

For many products, material flaws are not just cosmetic, but they can also affect product performance. The concept of zero defects means making absolutely 100 percent sure that customers never receive a flawed item. This type of quality inspection has become one of the most common machine vision applications. Vision systems are used to find missing material, chips, scratches, dents, misplaced markings, and other flaws. These inspections produce a secondary benefit by enabling manufacturers to eliminate defective pieces before wasting additional material and production time on them.

11

Summary and Conclusions

Reasons for Robots

Modern automation, by the use of multistation assembly machines and synchronized transfer lines, has greatly increased the rate of production and thereby lowered the costs of production for many items, both for industrial and consumer usage. In most examples, however, these machines are designed to do a limited range of operations for very long production runs. The more complex manufacturing tasks, which require flexibility, adaptability, and decision making, have necessarily been left to human beings. In addition, transfer lines and multistation assembly lines usually involve very large capital investment. This has, by itself, limited the opportunities where economic justification can be made. Figure 11.1 explains some relationships for several types of automation and the economic trade-offs that are involved.[1]

In the United States, it has been recognized that many jobs in manufacturing have been performed in unpleasant, unhealthy, or even dangerous environments. Congress has passed the Williams-Steiger Occupational Safety and Health Act of 1970, and recently revised Section Five of the Walsh-Healy Public Contracts Act. This action, along with the setting up of inspection systems, has gone a long way to correct adverse working environments. The operating part of the Act states:

"No part of such contract will be performed nor will any of the materials, supplies, articles, or equipment to be manufactured or fabricated

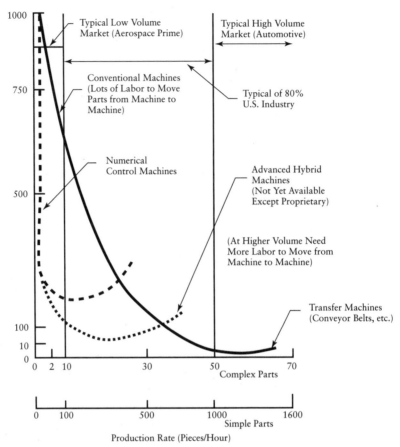

Source: "Revolutionaries Meet to Plot Strategies," Robert Haavind,
Computer Decisions. March, 1972; p. 16.

■ **Figure 11.1** *Economics of machine trade-offs.*

in any plants, factories, buildings, or surroundings or under working conditions which are unsanitary, or hazardous, or dangerous to the health and safety of employees engaged in the performance of said contract. Compliance with the safety, sanitary and factory inspection laws of the state in which the work or part thereof is to be performed shall be *prima facie* evidence of compliance. . . ."

In the United States, the rate of rise in wages has increased more than 70 percent over the last 10 years while productivity, the output per man-hour, has leveled off at about 2 percent per year. All of the principal competitors of the United States have had much greater increases in their productivity, ranging as high as 14 percent per year for Japan. Wages, on the other hand, have remained considerably below the high levels set in the United States.[1]

As a result of these pressures, the interest in industrial robots has greatly increased in all industrialized countries. These devices have been applied to many tasks not suited for human operators, and allow automation on a subsystem level where large-scale transfer lines and multistation assembly machines cannot be justified.

Robot Application Comparisons

It is important that each possible robot application be analyzed not only for its immediate potential savings, as illustrated by examples in this book, but that the sometimes hidden costs of maintenance, unused flexibility, space requirements, and higher quality of parts required also be factored into the considerations.

The newness and novelty of having an industrial robot, which can be programmed and directed to do one's unquestioned bidding, should not mesmerize one into overlooking the even greater savings that may be possible by employing fixed automation. Production rate and production volume will usually be the deciding factors. Table 11.1 lists some of the more dominant guidelines for successful application of an industrial robot.

Table 11.1 *Advantages of Manual Operation, Fixed Automation, and Industrial Robots for Increasing Productivity*

	Manual Operator	Industrial Robot	Fixed Automation
Capital Outlay	Lowest	Medium for Low Volume Runs	Highest for Low Volume Runs
Flexibility	Highest	Medium to High	Lowest
Speed	Lowest	Low to Medium	Highest
Toleration of Adverse Environment	Low	High	High
Space Requirements	Low	Medium	High
Mobility	High	Medium	Low
Toleration of Repetitive Tasks	Low	High	High
Quality of Parts Required	Lower	Higher	Higher
Quality of Parts Delivered	Lower	Higher	Higher
Maintenance Costs	Lower	Higher	Higher
Leadtime Required	Low	Low	High

Along with these general guidelines, it would be well to refer to the following checklist so that some obscure factor may not be overlooked in securing a successful robot application.

Checklist[2]

1. Does the application technically and economically warrant a robot installation?
2. Is it expected that future production runs and handling operations will need robot handling?
3. Is the number of operations for the robot so large that it might be better to have several simple robots working as a team?
4. Does the flexibility of future production increase if a number of simple robots are used?
5. Is special automatic equipment needed to serve the robot or is it better to use a robot that can search for objects and/or palletize?
6. Will the product have to be redesigned for robot handling?
7. Does fixed automation offer greater savings?
8. What is weight and geometry of part to be handled, now and in the future?
9. What accuracy is required in part positioning, both for loading and unloading?
10. What must the hand be like? Can a standard hand be used with minor alterations (weight, clamping force, size)? Can other types be used (e.g., suction pads or magnets)? Must a sensor be installed in the hand; what connections are required and where?
11. What type, if any, power actuators are required? It is necessary to force the part loose? What force is required? What power and movement characteristics, rate of acceleration, and rate of deceleration can be tolerated? What cycle time is required?
12. What type of power supply is needed (electrical, compressed air, or hydraulic power)? What capacity?
13. What type of control unit is required? Programmability? External communications? Is it better to use a minicomputer central control of several robots?
14. What are operating requirements (service, repair, and spare parts)? Is it advisable to have a standby robot? What is the economic and technical length of life? How easy is it to clear jams and get back

into production? What are emergency provisions? What is top speed? Is accuracy sacrificed? What are safety arrangements? Is fine adjustment easily done?

15. Is there sufficient space? What about floor loading? Can the robot be easily moved between different work stations?

Robot Applications

Following is a partial listing of some of the more successful robot applications gleaned from the literature:

- Brick handling
- Loading and unloading of presses
- Drop forging
- Paint spraying
- Die casting
- Glass handling
- Spot welding
- Assembly operations
- Testing high-voltage capacitors
- Shot peening
- Ceramic handling
- Heat treatment
- Machine tool servicing
- Feeding forming machines
- Feeding grinding machines
- Loading and unloading copy lathes
- Palletizing
- Surface coating
- Booking tire treads
- Handling glass
- Dismantling explosives
- Packaging
- Warehousing

Future Developments

Driscoll has separated robots into three generations.[3] "The first generation was robots of 1975, which performed isolated motor functions of a human being. For example, they load, unload, transfer, stack, handle, and palletize.

"Generation 2 machines will be man substitutes in a more complex sense. They will perform isolated functions, some perceptual motor functions, of the blue-collar worker.

"The third generation robots will take advantage of the tremendous effort being made in the development of artificial intelligence. They will approach the true functions of the living organism. This is the central theme of cybernetics: to use the same logic and mathematics to describe both men and machines. This level of robot was to McCullock a prescription for a system that until recent advances in artificial intelligence, could only be achieved by the growth of living cells."[4]

As the second and third generation of robots arrive on the scene, the same kind of system analysis, economic justification, and comparison with other forms of automation described in this book must be made. Certain unique features of these robots are sure to make them singular solutions to special problems. But the industrial robot, like its earlier counterpart, the digital computer, will only achieve its true place in the hierarchy of automation, when it is properly applied to support man in his ever increasing search for productivity.[5]

Recent developments in the Mars Lander robots Spirit and Opportunity, robots used to perform medical operations where a steady hand is needed, or a new robot being used to pick lettuce in the fields at Kalamazoo (using machine vision to determine ripeness), all challenge man's imagination and creativity to make use of this valuable servant. Work is needed to explore further applications of machine vision, expanding techniques for utilizing various parts of the spectrum, especially where the human eye is not quite so sensitive. We need to explore all the other human senses—hearing, touch, taste, and smell. The future of robotics could use all of these if proper sensors were available.

It has been predicted, however, that flexible automation (which includes multipurpose robots and programmable equipment of all types) will *not* outsell fixed automation, which is the older and less sophisticated of the two types. One expert, Harold N. Bogart, Director of the Manufacturing Development Office, Engineering & Manufacturing Staff, Ford Motor Co., said: "I don't think programmable robots or other flexible devices will ever replace fixed automation in our industry. They'll grow in application. But in my opinion, they're often too expensive and versatile for the simple applications we have in mind, and won't take over any operations that can be handled by fixed automation."[6]

References

1. Kornfeld JP, Magad EL. Robotry and Automation—Key to International Competition. *Proceedings of Second International Symposium on Industrial Robots.* 41.

2. Lundstrom G, Lundstrom L, Arnstrom A, Rooks B. *Industrial Robots—A Survey.* Bedford England: Pub. International Fluidics Services, Ltd.; 1972:14.

3. Driscoll LC. Blue Collar Robots—A Technology Forecast. *Proceedings of Second International Symposium on Industrial Robots.* Chicago: IIT Research Institute; May 16–18, 1972:197–201.

4. Driscoll LC. Second and Third Generation of Industrial Robots. *Proceedings of the First National Conference on Remotely Manned Systems.* California Institute of Technology; September 13–15, 1972.

5. Wight O. *The Executive's New Computer.* Reston, VA: Reston Publishing Co.; 1972.

6. Bogart HN. Speech at Society of Automotive Engineers' Congress and Exposition. Detroit, Michigan; 1972.

Appendix

A Survey of
Industrial Robots

ABB Inc., 2487 S. Commerce Dr., New Berlin, WI 53151, 262-785-3400 or 888-785-3904. Contact: Ann Smith, Dir. of Mktg.

Action Machinery Co., One Vulcan Dr., P.O. Box 307, Helena, AL 35080. Contact: David LaRussa, 205-663-0814.

Adept Technology, 3011 Triad Dr., Livermore, CA 94551. Contact: John Dulchinos, 925-245-3400.

Alvey Systems Inc., 9301 Olive Blvd., St. Louis, MO 63132. Contact: Virgil Beer, Mktg. Mgr., 314-872-5719.

American Machine and Foundry, Versatran Div., 26422 Groesbeck Hwy., Warren, MI 48089.

AMI Inc., US Route 22, Whitehouse, NJ 08888.

Anorad Corp., 110 Oser Ave., Hauppauge, NY 11788. Contact: Michael Backman, Dir. of Mktg., 631-231-1995.

Animatics Corp., 3050 Tasman Dr., Santa Clara, CA 95054, 408-748-8721.

Taken in January 2004; sources located on the Internet.

Antenen Research, 4300 Dues Dr., Cincinnati, OH 45246. Contact: Marshall Burke, Mktg. Dir., 513-860-8807.

Autoplace, 1401 East Fourteen Mile Rd., Troy, MI 48084.

ATS-Automation Tooling Systems, Inc., 250 Royal Oak Rd., Cambridge, Ont. Contact: James Beretta, 519-653-6500.

Bergandi Machine Company, 1520 Adelia Ave., S. El Monte, CA 91733.

Barret Technologies Inc., 139 Main St., Kendall Square, Cambridge, MA 02142. Contact: Kellie Browne, 617-252-9000.

Bosch Rexroth Corp., 816 E. Third St., Buchanan, MI 49107. Contact: Kevin Gingerich, 616-695-5363.

Brenton Engineering Co., 4750 County Rd. 13 NE, Alexandria, MN 56308. Contact: Karen Kielmeyer, 320-852-7705.

Banner Welder, Inc., 18200 Fulton Dr., Germantown, WI 53022. Contact: Robert Kerr, 414-253-2900.

Burch Controls, P.O. Box 24, Kalamazoo, MI 49004.

Castle Industrial Ergonomics, Inc., 302 South Spring St., Manchester, TN 37355. Contact: J. Keith James, 931-728-7733.

Cloos Robotic Welding Inc., 911 Albion Ave., Schaumburg, IL 60193, 847-923-9989.

Cobotics Inc., 2506 Gross Point Rd., Evanston, IL 60201, 847-425-1200.

Columbia/Okura LLC, 301 Grove St., Vancouver, WA 98661, 360-735-1952.

Comau Pico, 21000 Telegraph Rd., Southfield, MI 48034, 248-353-8888.

Commotion Technology, World Trade Center, Suite 250L, San Francisco, CA 94111. Contact: Marcus Ruark, 415-391-9212.

Creative Automation Inc., 4843 Runway Blvd., Ann Arbor, MI 48108, 734-930-0050.

Custom Control Panels Inc., 1686 Mattawa Ave., Mississauga, Ont. L4X1K2 Canada, 905-279-6271.

Cybo Robots, 2040 Production Dr., Indianapolis, IN 46241. Contact: Ron Smith, 317-484-2926.

Centerline Automation Services Inc., 820 Boston Turnpike Rd., Shrewsbury, MA 01545. Contact: Nathan Pelis, 508-845-8697.

Crown Hollander, Inc., 311 Enford Rd., Richmond Hill (Toronto), Ont. L4C3E9 Canada. Contact: Harold H. Holander, 905-884-1263.

Diahen Inc., 5311 W.T. Harris Blvd. West, Charlotte, NC 28269, 704-597-8240.

Denso Robotics, 3900 Via Oro Ave., Long Beach, CA 90810. Contact: Peter Cavallo, 310-513-7343 or 800-222-6352.

Dispense Works, 3980 Albany St., McHenry, IL 60050. Contact: Emil Cindric, 815-363-3524.

Epson America, Factory Automation/Robotics, 18300 Central Ave., Carson, CA 90746. Contact: John C. Clark, 562-290-5910.

Erowa Technology, Inc., 2535 Clearbrook Dr., Arlington Heights, IL 60005. Contact: Skip Thompson, 847-290-0295.

FANUC Robotics North America, 3900 Hamlin Rd., Rochester Hills, MI 48309. Contact: Kathy Powell, 248-377-7570.

Genesis Systems Group, 8900 Harrison St., Davenport, IA 52806. Contact: Mike Jacobsen, Mktg. Mgr., 563-445-5600.

Gregors Location Services, Box 350, Michigan Ave., Howell, MI 48844. Contact: John Gregor, 517-548-7111.

Hyundai Engine and Machinery, Factory Automation and Robots, Cheonha-Dong, Dong-Ku, Ulsan, Korea. Contact: S.G. Chun, Director, 82-52-230-7900.

HRI, Inc., Machining Specialists, P.O. Box 867, Salem, NH 03079. Contact: Krysten L. Magoon, 603-894-5662.

Innovative Robotic Solutions, Inc., 2910 Scott Blvd., Santa Clara, CA 95054. Contact: Dick Dexter, 408-919-1801.

International Engineering Solutions, Inc., P.O. Box 518, North Jackson, OH 44451. Contact: David Wigal, 330-538-0068.

I&J Fisnar, Inc., 2-07 Banta Pl., Fair Lawn, NJ 07410. Contact: Warren Noen, Mktg., 201-796-1477.

Industrial Control Repair, 12934 Ten Mile Rd., Warren, MI 48089. Contact: Glen Dantes, 586-757-8072.

Intelligent Actuator, Inc., W. 237th St., Torrance, CA 90505, 310-891-6015 or 800-736-1712.

Kawasaki Robotics USA Inc., 28059 Center Oak Ct., Wixom, MI 48393. Contact: Kathy Gulewich, Mktg., 248-305-7610.

KC Robotics, 9000 Lesaint Dr., Fairfield, OH 45014. Contact: Bobby Griffin, 513-860-4442.

Kuka Robotics, 6600 Center Dr., Sterling Heights, MI 48312. Contact: Yarek Niedbaia, 866-873-5852.

Lincoln Electric Co., 22801 St. Clair St., Cleveland, OH 44117, 216-481-8100.

Mekanize Robots & Engineering Inc., 1420 Quail Lake Loop, Colorado Springs, CO 80906. Contact: Jeff Clear, 719-598-3555 or 800-482-1834.

Materials Handling Enterprises, Box 8348, Erie, PA 16505. Contact: David P. Snell, 814-454-6396.

Mentor AGVS-Formtek Cleveland Inc., 26565 Miles Rd., No. 200, Cleveland, OH. 216-292-6300 or 800-631-0520.

Mitsubishi Electric Automation Inc., 500 Corporate Woods Pkwy., Vernon Hills, IL 60061, 847-4778-2282.

Motoman Inc., 805 Liberty La., West Carrollton, OH 45449. Contact: Sally Fairchild, Mktg. Mgr., 937-847-3288.

MRM Inc., 22777 Heslip Dr., Novi, MI 48375. Contact: Mike Brotz, 248-348-6900.

Machine Control Services, 1999 Theresa La., Moody, TX 76557, 254-853-3925.

Nachi Robotics Systems Inc., 22285 Roethel Dr., Novi, MI 48375. Contact: Karen Lewis, 248-305-6545.

Nimbl Inc., 105 A-8 Commerce Circle, Madison, AL 35758. Contact: Rory Flemmer, Pres., 256-464-5568.

Opus Automation Inc., 844-A Trilliam Dr., Kitchener, Ont. N2R1J9 Canada, 519-893-9331.

Pacific Robotics Inc., 7966-C Arjons Dr., San Diego, CA 92126. Contact: Paul Sager, Pres., 858-408-0707.

Panasonic Factory Automation, 1711 Randall Rd., Elgin, IL 60123. Contact: Brian Knier, Mktg. Mgr., 858-304-5860.

Par Systems Inc., 899 Highway 96 West, Shoreview, MN 55126, 651-484-7261.

PRI Robotics, 2440 Fernbrook La., Minneapolis, MN 55447. Contact: Lynn Swanson, 763-559-2115.

Reis Robotics USA, Inc., 1320 Holmes Rd., Elgin, IL 60123. Contact: Tim Pittenger, 847-741-9500.

Radix Controls, 2105 Fasan Dr., P.O. Box 1, Oldcastle, Ont. N0R1L0 Canada. Contact: Shelly Fellows, 519-737-1012.

Rixan Associates, 7560 Paragon Rd., Dayton, OH 45459. Contact: Stephen Harris, Pres., 937-438-3005.

Robot Company, The, 300 Ryan Rd., Seville, OH 44273. Contact: David C. Beard, Sales Mgr., 330-769-4900.

Robert I. Robotics, Inc., 343 Sovereign Rd., London, Ont. N6M1A6 Canada. Contact: Robert Isaac, Mgr., 519-453-4165.

Robotics Inc., 2421 Route 9, Ballston Spa, NY 12020. Contact: Joe Morgan, Mktg. Mgr., 518-899-4211 or 800-876-2684.

Robotic Concepts Inc., 1425 Alloy Pkwy., Highland, MI 48356. Used robots and robot systems, 248-889-0064.

Robots Dot Com Inc., 2995A South Mooreland Rd., New Berlin, WI 53151. Contact: Glen R. Blok, 262-789-1366.

RP Automation Inc., 56 Dodgingtown Rd., Bethel, CT 06801. Contact: Ben Clark, 203-790-5787.

Robotic Workspace Technologies Inc., 16266 San Carlos Blvd., Ft. Myers, FL 33908. Contact: Robert Ballard, 239-466-0488.

Roboprobe Technologies Inc., P.O. Box 1037, Palatine, IL 60078, 847-934-5567.

Sig Pack Systems, U.S. Division, 2401 Brentwood Rd., Raleigh, NC 27604. Contact: Brad Baker, 919-877-2015.

Staubli Corp., P.O. Box 189, Duncan, SC 29334. Contact: Jeff Parker, 864-433-1980.

S.J. Industries Inc., 3235 Twain Circle, Brunswick, OH 44212. Contact: Steve J. Wasylko, 330-225-7969.

Specialty Equipment—Fillers, Conveyors, Robots, 1221 Adkins, Houston, TX 77055, 713-467-1818.

Sundstrand Machine Tool, Belvidere, IL 61008.

Syncro-Trans Corp., 19250 Plymouth Rd., Detroit, MI 48228.

Thermo CRS (Formerly CRS Robotics), 5344 John Lucas Dr., Burlington, Ont. L7L6A6 Canada. Contact: Kevin Keras, 905-332-2000.

Unimation Inc., Shelton Rock La., Danbury, CT 06810.

U.S. Digital, 11100 NE 34th Circle, Vancouver, WA 98682, 360-260-2468.

VSI Automation Assembly, 2700 Auburn Ct., Auburn Hills, MI 48326. Contact: Robert Mineau, 248-853-0555.

Weiser/Robodyne Corp., 949 Bonifant, Silver Spring, MD 20910.

Weldotron Corp., 1532 S. Washington Ave., Piscataway, NJ 08854.

Wes-Tech Automation Inc., 720 Dartmouth Lane, Buffalo Grove, IL 60089. Contact: Greg Olcnila, 847-541-5070.

Wickes Corporation, 515 Washington, Saginaw, MI 48607.

Wittmann Robot & Automation Systems Inc., 1 Technology Park, Torrington, CT 06790. Contact: Michael Wittmann, 860-496-9603.

Yamaha Robotics, P.O. Box 937, Edgemont, PA 19028. Contact: Doug Dalgliesh, 610-325-9940.

Yashin America, 35 Kenney Dr., Cranston, RI 02920, 401-463-1800.

Index

208

off-highway farm tractors, 121–122,
122
off-line programming (OLP), 76
off-road equipment industry, 118–120,
119, 120
OSHA. *See* Occupational Safety and
Health Act of 1970
Otani, Ernest M., 68
output per man hour, economic productivity and, **10**

pace of work, 166
packaging industry, 106–107, **107**
paint. *See* automotive paint shops, 68
palletizing systems, 100–102, **102,**
106–107, **107,** 108–109, **109,**
144–145, 146–147
Paul, Jim, 48
payback period, 153–155
perimeter guarding, 183
personal care industry, 100–102, **102**
physical working conditions, 165
plastic injection molding, 141–143, **143**
plastic molders, 116–117, **117**
polishing. See Kuntz Electroplating
Inc.; wheel polishing process
automation
power generation industry, 135–137,
136, 137
Presence Sensing Device Initiation
(PSDI), 185
present value of future earnings,
154–155, **155**
print verification systems, 188
process control, 60
productivity. *See* economic productivity
program length requirements, 25
programming requirements, 25, 60, 65
PushCorp Inc., 75, 79–80

quality, 32
flaw detection systems and, 189

RADCO Industries, 44, 46–49
reach and volume of arm movement, 26
reasons for use of robots, 191–193, **192**
rebuilt or redeployed robots, safety
issues of, 176–177
redeployed robots, safety issues and,
176–177

retrofits and grandfathering, safety
issues of, 175
return on investment, 154
Rimrock Automation Inc., 173
risk assessment, safety issues and, 170,
182
Robocad open system environment
(ROSE), 78–79
robot controller software (RCS), 78
robotic guidance systems, 189
Robotic Industries Association (RIA),
safety standards of, 169–172
robotic vision systems, 38–42, **40**
application time requirements for,
38–39
calibration of, 39
code reading systems in, 188
complexity requirements for, 39–40
dimensional gauging systems and,
189
ease of programming in, 40–41
errors in, 41–42
flaw detection systems and, 189
*Impact of Machine Vision on the
Robotics Industry* and, 2,
187–189
laser-based, 51
machine vision and, 187–188
print verification systems in, 188
robotic guidance systems and, 189
teach pendant, **41,** 41
Robotics Industries Association (RIA),
63
robots
fixed automation vs., 15–21
human motion times vs., 23–25, **24**
liquid level monovial assembly and,
15–19, **16, 17, 18**
tending die-casting machines, 19–21,
20
roller bearings, 92–94, **94**
Rossum's Universal Robots (RUR), 1

safety, 65–66, 169–186
ANSI/RIA R15.06-1999 standard
for, 172–177
compliance with codes, 171
current issues in, 174–177
design process and, 182–184
FAQs on, 177–180

upgrading of worker skills, 165
upkeep. *See* maintenance and operating
    costs
urethane dispensing system, 121–122,
    122

Versatran, 2–3
Vision Systems Design, 38
vision. *See* robotic vision systems
visLOC systems, 38, 41

wage averages for industrial workers,
    12, **12**

Walsh-Healy Public Contracts Act,
    191–192
weight lifting requirements, 25
welding, seam tracking systems for,
    50–53, **52**, **53**
wheel deburring systems, 138–140, **139**,
    140
wheel polishing process automation,
    74–81, **81**
Williams-Steiger Occupational Safety
    and Health Act of 1970,
    161–163, 191

About the Author

HARRY COLESTOCK is an electrical engineer who designed and built automation systems for over 40 years. A specialist in robot productivity improvement, he has analyzed robotics for hundreds of companies in the auto industry as well as other high-volume industries. As chief engineer of Advanced Technology and Testing at Ingersoll-Rand, Mr. Colestock led the effort in the design and installation of some of the most advanced robotic systems in the world. His last job, before retiring, was the design and installation of an entire assembly and test line for the automatic assembly and testing of the six cylinder engine for General Motors' Holden division in Melbourne, Australia. He now lives in Petoskey, Michigan.